EXOTIC PROPERTIES OF SUPERFLUID ³He

SERIES IN MODERN CONDENSED MATTER PHYSICS

ISSN: 2010-2119
Editors-in-charge: I. Dzyaloshinski and Yu Lu

Published

Series in
Modern
Condensed
Matter
Physics

Vol. 1

EXOTIC PROPERTIES
OF SUPERFLUID ^{3}He

G. E. Volovik

Landau Institute for Theoretical Physics, Moscow

World Scientific

Singapore • New Jersey • London • Hong Kong

Published by

World Scientific Publishing Co. Pte. Ltd.

5 Toh Tuck Link, Singapore 596224

USA office: 27 Warren Street, Suite 401-402, Hackensack, NJ 07601

UK office: 57 Shelton Street, Covent Garden, London WC2H 9HE

British Library Cataloguing-in-Publication Data
A catalogue record for this book is available from the British Library.

Series in Modern Condensed Matter Physics — Vol. 1
EXOTIC PROPERTIES OF SUPERFLUID ³He

ISBN-13 978-981-02-0705-2
ISBN-10 981-02-0705-0
ISBN-13 978-981-02-0706-9 (pbk)
ISBN-10 981-02-0706-9 (pbk)

Foreword

John Bardeen once compared condensed matter physics with a huge deep-rooted tree. Some of its shoots may dry up, but there are always new branches which will grow, prosper and blossom. In the last decades we have witnessed so many exciting developments in condensed matter physics: superfluidity in ^{3}He, superconductivity in organic compounds and oxides, integer and fractional quantum Hall effect, to name just a few examples. At the same time, a number of new concepts like symmetry breaking, coherence, scaling and fractional statistics as well as new powerful techniques like renormalization group, topological methods and other field-theoretical tools, have appeared. Such rapid development of modern condensed matter physics calls for a wide activity in publications at different levels. In the current literature there are plenty of conference proceedings and some comprehensive review volumes accessible mainly to experts. On the other hand, the time is not ripe yet for many of these new developments to be included in standard textbooks.

The aim of this series is to provide an introduction to new areas of research in condensed matter physics, mainly addressing a young readership: graduate students and beginning research workers. The basic format is an independent volume by an expert author of around 150 pages each. These volumes could be lecture notes or short monographs of a lighter style. The emphasis will be on new physical concepts and interconnections with other phenomena and fields, while the corresponding mathematical tools will be used whenever necessary. In selecting topics we tried to pick those of broader current scientific interest but avoiding too "hot" ones so that these volumes can stay on the shelf for some time.

We would like to express our sincere thanks to the authors of individual volumes who kindly agreed to take up this difficult project. We are sure that every active researcher will deeply appreciate the authors' willingness to spend part of their precious research time for this educational purpose. Hopefully, this will be a rewarding experience, as bright new minds will come to the field, encouraged and well prepared by their predecessors. We would highly appreciate comments and suggestions from readers and potential authors.

Igor Dzyaloshinkii
Stig O. Lundqvist
Yu Lu

Contents

Contents

EXOTIC PROPERTIES OF SUPERFLUID ³He

1
Introduction

Liquid helium-4 and liquid helium-3 and their solutions are quantum liquids since the quantum nature of physical laws directly governs the properties of these condensed matter substances at temperatures of several kelvins. These are the only liquids which do not solidify even at absolute zero temperature due to large zero-point motion of atoms. Though the atoms ^{3}He and ^{4}He are indistinguishable on a chemical level due to the identical structure of electronic shells, the principal difference in the properties of their nuclei (the ^{4}He nucleus has zero spin, while the nuclear spin of the ^{3}He atom is $\frac{1}{2}$) becomes important at low temperatures when the thermal noise is too small to suppress the manifestation of quantum effects. At a temperature of several kelvins the difference in quantum statistics of the nuclei of helium isotopes becomes crucial: the ensemble of the ^{4}He atoms forms the quantum Bose liquid while the ensemble of the ^{3}He atoms forms the quantum Fermi liquid.

This results in enormous, by three orders of magnitude, difference in the temperature T_c of phase transition into the superfluid state: $T_c = 2.2$ K for ^{4}He (λ-point) and $T_c = 2.7$ mK for ^{3}He. This difference takes place because the superfluid transition is quite a natural process for the Bose system, but not for Fermi systems. The former displays the phenomenon of the Bose condensation: formation of the highly coherent ensemble in which

1

all the atoms are in the same quantum state. The Bose condensation is the basis of the phenomenon of superfluidity which manifests itself in the rigid correlated motion of the Bose condensate without any dissipation.

Superfluidity of the Fermi liquid helium-3 is essentially more complicated than that of the Bose liquid helium-4, because to form the superfluid Bose condensate the atoms of ^{3}He must first form bosons. The mechanism of the formation of bosons in a Fermi liquid is known in metals, for electronic liquid in superconductors. This is the pairing of electrons with the formation of the so-called Cooper pair with integer spin. The ensemble of the bosonic Cooper pairs forms the Bose condensate which is responsible for the superfluidity of electric charge, i.e., superconductivity.

In conventional superconductors the spin of Cooper pair is zero, $S = 0$, and this reminds us of the spinless atom ^{4}He resulting in many common features shared by the superfluid ^{4}He with a conventional superconductor. Both bosons also have no internal orbital angular momentum, $L = 0$, i.e., the Cooper pair in conventional superconductors is spherically symmetric in the same manner as the ^{4}He atom. On the characteristic energy scale $E \ll T_c$ both particles may be considered as elementary particles without any internal degrees of freedom. Since bosons comprising the Bose condensate in superfluid ^{4}He and conventional superconductors, atoms of helium-4 and Cooper pairs, are structureless units on an energy scale relevant for superfluidity and superconductivity, the vacuum of these substances, the ground state, is described by only a one-component complex order parameter $\psi =| \psi | \exp(i\Phi)$, the macroscopic wave function of coherent motion of structureless particles. This is an example of the scalar field theory in condensed matter physics.

While the modulus of ψ in the equilibrium state of ^{4}He or a superconductor is fixed by external conditions (temperature, pressure, etc.), the phase Φ of the order parameter is arbitrary with different Φ's coresponding to distinguishable equilibrium states. This phase Φ thus parametrizes the continuously degenerate equilibrium states of superfluid helium-4. The distinguishability of the phase means that the gauge symmetry $U(1)$ is spontaneously broken in superfluids and superconductors. The broken gauge symmetry is responsible for the main features of the superfluid state of

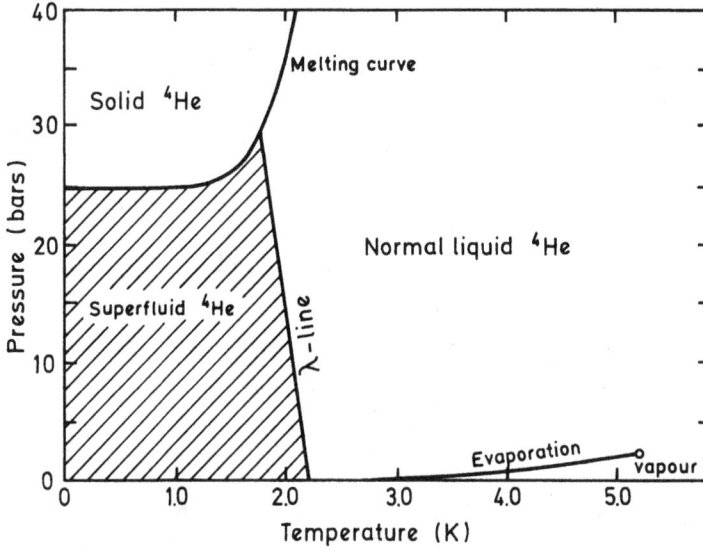

(a)

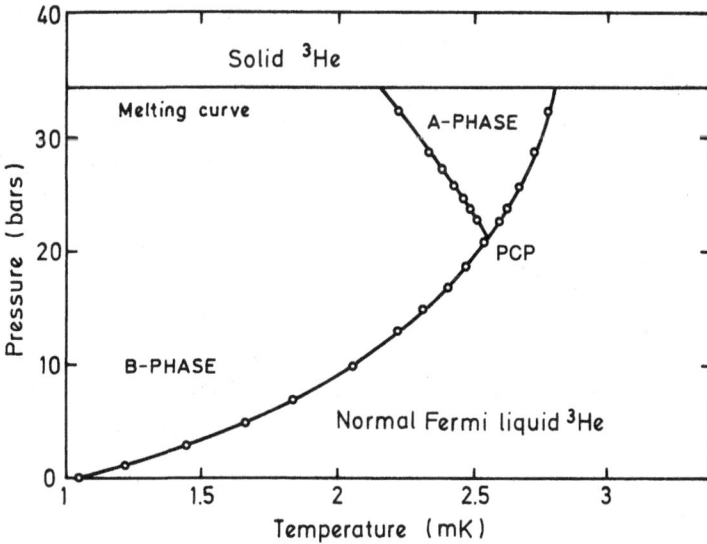

(b)

Fig. 1.1. Pressure vs. temperature phase diagram for ^{4}He (a) and for ^{3}He (b). Two chemically indistinguishable liquids have a three orders of magnitude difference in the superfluid transition temperature.

liquid helium-4 below T_c as compared with the normal state above T_c:

1) Frictionless motion of liquid with the irrotational superfluid velocity

$$\vec{v}_s = (\hbar/m_4)\ \vec{\nabla}\Phi\ ,\tag{1.1}$$

here m_4 is the mass of the ⁴He atom.

2) New sound wave collective mode related to the oscillations of the new soft hydrodynamical variable Φ.

3) Existence of topologically stable objects, quantized vortices, with the integer winding number of the phase Φ around the vortex in terms of 2π.

4) Josephson effect of quantum coherence; etc. ... .

In superfluid ³He the elementary particles which form the Bose condensate have nontrivial internal structure, giving rise to several additional degrees of freedom even on energy scale $E \ll T_c$. This results from the nonzero spin of the Cooper pair, $S = 1$, and from the nonzero angular momentum of the orbital motion of the Cooper pair about its center of mass, $L = 1$. In the highly coherent Bose condensate the spin and orbital momenta of Cooper pairs are strongly correlated, resulting in magnetic and liquid-crystal-like ordering. This means that in addition to the broken gauge symmetry the symmetry $SO_3^{(S)}$ under spin rotations and the symmetry $SO_3^{(L)}$ under orbital rotations are spontaneously broken in superfluid ³He.

Due to the 3×3 degrees of freedom, $(M^S = -1,0,+1) \times (M^L = -1,0,+1)$, where M^S and M^L are the spin and orbital momentum projections respectively, the states of superfluid ³He are described by macroscopic wave functions with nine complex, and therefore 18 real, amplitudes. The complicated structure of the Ginzburg-Landau functional and multicomponent nature of the order parameter result in the essentially new features of superfluid helium-3 as compared with helium-4. First, instead of only one superfluid phase of helium-4 there may be several different phases of superfluid helium-3 with essentially different internal symmetries of their vacuum states, and therefore with different physical properties. Three of them are realized in different regions of pressure and magnetic field and are under extensive experimental investigation. These are:

1) The quasi-isotropic phase ^{3}He-B, which corresponds to Cooper pairing into a state with zero total angular momentum, $S = 1$, $L = 1$, $J = 0$, where $\vec{J} = \vec{L} + \vec{S}$;

2) The anisotropic ^{3}He-A. The Cooper pair in this state has nonzero projection of the orbital momentum $M^L = 1$ on some axis $\hat{l}$. This unit vector $\hat{l}$ is simultaneously an axis of spontaneous orbital anisotropy of this liquid and the direction of its spontaneous ferromagnetic moment. The arrangement of the nuclear spins of the Cooper pair in the A-phase state also has some preferred axis $\hat{d}$, however this is not an axis of the net magnetic moment, but of magnetic anisotropy like in collinear antiferromagnets. Formally this unit vector $\hat{d}$ indicates the axis of quantization of the spin S of the Cooper pair: the projection M^S on this axis is zero, $M^S = 0$.

So the superfluid ^{3}He-A shares the properties of orbital ferromagnet with spontaneous magnetization along $\hat{l}$, uniaxial liquid crystal with anisotropy axis along $\hat{l}$, and spin antiferromagnet with magnetic anisotropy axis along $\hat{d}$.

3) The ^{3}He-A$_1$ phase exists only in the presence of a magnetic field. It has $M^L = 1$ and therefore the orbital ferromagnetism like in A-phase, and in addition exhibits ferromagnetic ordering of nuclear spins, i.e., $M^S = 1$. The axes of orbital and nuclear spontaneous magnetization are not necessarily parallel.

Some other superfluid phases may be realized under special conditions: i) in confined geometry, ii) in superfluid films, iii) in cores of quantized vortices or other topological objects, iv) on the surface of containers, v) in the tiny temperature region near T_c, etc.

An important feature of the superfluid ^{3}He phases is the high degeneracy of its equilibrium states, which is directly related to more extended symmetry breaking. Instead of the one-dimensional space of the degenerate equilibrium states in superfluid ^{4}He, distinguished by the phase Φ of the Bose condensate, the manifold of the degenerate states is 4-dimensional in the B-phase and 5-dimensional in the A-phase.

This results in a large number of collective modes. Each phase has at least 18 modes, of oscillations of 9 complex components of the order parameter near one of the degenerate equilibrium states. Some of these modes are soft, i.e., have no gap in the spectrum $\omega(\vec{q})$ like the fourth sound in superfluid ^{4}He. These are the so-called Goldstone bosons resulting from the continuous degeneracy of the equilibrium states, and therefore from the broken continuous symmetry.

In ^{3}He-A there are 5 Goldstone bosons, in accordance with the dimension of the manifold of degenerate states, in addition to the fourth sound there are two polarizations of spin waves like in antiferromagnets and two polarizations of the orbital waves. The latter are the propagating oscillations of the common axis $\hat{l}$ of the orbital ferromagnetism and liquid-crystal anisotropy, while the former are the propagating oscillations of the $\hat{d}$ vector. Also, 4 Goldstone bosons are present in ^{3}He-B.

As a result of the additional soft modes, the hydrodynamics (dynamics of the soft variables) of the superfluid phases of ^{3}He is much more complicated than in ^{4}He and contains a lot of new effects, which couple different magnetic, liquid-crystal and superfluid properties. It is important that the coupling between different hydrodynamical variables is not simple, due to another new phenomenon of the superfluid ^{3}He known as broken relative symmetry. In superfluid ^{3}He-A this phenomenon intrinsically couples the liquid-crystal-like and superfluid properties in such a manner that the superflow becomes nonpotential in the presence of textures, the nonuniform distributions of the (liquid-crystal) order parameter ($\hat{l}$). This introduces a new class of superfluids, which differs from the "classical" superfluids where the superflow is irrotational.

The high degeneracy of the superfluid states of ^{3}He leads also to a number of textures, inhomogeneous distributions of the order parameter. Some of them, like the quantized vortices in superfluid ^{4}He, prove to be extremely stable due to the conserved charges induced by their topological properties. In addition to many different types of quantized vortices there are domain walls, solitons, particle-like topological defects, disclinations, hedgehogs, monopoles, various defects on the surface of the container and on the phase boundary between ^{3}He-A and ^{3}He-B (boojums), solitons terminating on the disclinations, etc. ... .

These inhomogeneous structures differ not only by their topological charges, i.e., conserved topological numbers, but also by their symmetry.

The latter is illustrated by the observed phase transition between two types of quantized vortices in superfluid ^{3}He-B: they are topologically equivalent, i.e., they have the same winding number of the condensate phase Φ around the vortex axis, but differ by the symmetry properties of their cores. Both vortices exemplify the spontaneous breaking of symmetry in topological objects: the space parity is broken in the cores of both vortices, which leads to spontaneous electric polarization along the vortex axis. In addition, in one of the vortices the axial symmetry of the vortex core is broken, which leads to a new Goldstone boson propagating along the vortex axis. This soft mode has recently been observed in NMR experiments.

Superfluid phases of ^{3}He provide also examples of closed quantum field theory. There are three types of elementary excitations (elementary particles) of superfluid vacuum:

1) bosonic elementary particles – quanta of 18 collective modes, including the massless Goldstone bosons;

2) fermionic quasiparticles originating from the initial fermions, ^{3}He atoms, dressed by the Fermi-liquid interaction and modified by superfluid pair correlations, these are the so-called Bogoliubov excitations; and

3) particle-like topological objects which in the quasi-two-dimensional superfluid systems (films, interfaces, domain walls) may obey different quantum statistics: the Fermi statistics as well as Bose statistics and even the intermediate statistics, parastatistics.

The dynamics of the elementary particles interacting with each other and with the continuous order parameter field, which plays the part of gauge fields, constitutes a self-consistent quantum field theory (QFT). This QFT does not suffer from divergencies, as distinct from the quantum field theories in particle physics, since the structure of the vacuum states in superfluid phases of ^{3}He is completely known. There is no cutoff problem in the ultraviolet limit which plagues the relativistic quantum field theories, since it is known in principle what is beyond the cutoff parameters in ^{3}He.

In the low energy limit, when the symmetry of the fermi and bose quasiparticles of the superfluid phases of ^{3}He is enlarged to become "relativistic", the QFT in these phases is in many details similar to that in elementary particle physics. In some cases there is even one-to-one correspondence between some details, which leads to several identical effects described, perhaps, by different physical language.

For example the fermions in ^{3}He-A are massless chiral fermions, they obey the Weyl equation like neutrino or left and right electrons, while the fermions in ^{3}He-B are massive and obey the Dirac equation. Some components of the order parameter field in ^{3}He-A, the $\hat{l}$ texture, interact with the fermions in the same manner as the electromagnetic gauge field interacts with electrons. The propagating orbital waves thus correspond to the photons. Some other components of the order parameter represent gravitational field and weak-interaction gauge field: they are felt by the fermions as space curvature or as local $SU(2)$ gauge field. The corresponding propagating waves, clapping modes and spin-orbital waves, correspond to gravitons and W-bosons.

The similarity of the low energy dynamics leads to analogous phenomena in ^{3}He-A and high energy physics. So to describe a phenomenon in condensed matter, one should sometimes translate it into the language of particle physics, and vice versa. The examples are: 1) the chiral anomaly (nonconservation of the conserved charge due to its creation from the vacuum degrees of freedom) and 2) the zero-charge effect (logarithmical screening of charge due to dielectric polarization of vacuum). In the language of superfluids these are correspondingly 1) the momentum transfer from the coherent motion of the Bose condensate (motion of the superfluid component of liquid) to the normal component of liquid and 2) the logarithmical divergency of the gradient energy. In spite of the different language – the chiral charge corresponds to the linear momentum projection on the $\hat{l}$ vector – these phenomena are described by the same equations, which exhibit local gauge invariance and general covariance.

On the other hand the nature of some analogous phenomena in the helium-3 liquids does not coincide with the theoretical interpretations proposed in traditional models of field theories. For example the finite mass of the "W-boson" in ^{3}He-A does not originate from the Higgs mechanism as in the Weinberg-Salam theory. The mass of the W boson is exactly zero within the Bardeen-Cooper-Schrieffer (BCS) theory of superfluid ^{3}He due to some hidden symmetry of the BCS Hamiltonian. The mass becomes nonzero due to the small corrections to the BCS Hamiltonian when one goes beyond the BCS theory; these corrections violate the hidden symmetry. Incidentally just these corrections stabilize the ^{3}He-A state at high pressure, since the BCS weak-coupling theory gives for the A-phase higher energy than for the B-phase.

The other difference comes from the estimation of the cosmological term in the gravity theory. There are some difficulties in cosmology related to the calculated high value of the cosmological constant due to vacuum fluctuations, which gives a huge gravitational mass of the vacuum, 120 orders of magnitude larger than the experimental upper limit. The B-phase represents an example of the bimetric theory of gravitation, one metric tensor represents the equilibrium state of the vacuum, while the second one is dynamical and represents the gravity. The collective modes which give rise to the gravitation field are the so-called real squashing modes intensively studied in ultrasonic experiments. The existence of two metric tensors modifies the cosmological term in the Einstein theory of gravity. Though the cosmological constant proves to be rather high due to vacuum fluctuations, this does not produce the gravitational mass of the vacuum and therefore does not lead to problems in cosmology.

The topological objects in superfluid ^{3}He are also in many details similar to those of relativistic QFT; the main difference is that in superfluid ^{3}He they may in principle be observed and investigated while in elementary particle physics they are still hypothetical objects. There are a lot of examples of the topological confinement of defects in superfluid ^{3}He, which reminds one of the quark confinement in particle physics. The point defects, the $\hat{l}$ field hedgehogs, which are analogous to the Dirac magnetic monopoles, are confined by the segment of the quantized vortex, which are thus analogous to the Dirac string. Other monopoles without strings, the $\hat{d}$ hedgehogs, are equivalent to the t'Hooft-Polyakov magnetic monopoles. The half-quantum vortices in ^{3}He-A and the disclinations in ^{3}He-B are confined by the solitons. The latter object, the soliton, terminating on string, was recently observed in rotating cryostat.

The quantized vortices extensively investigated in superfluid ^{3}He-B, mainly by NMR methods, are the counterparts of cosmic strings that are believed to have precipitated in the phase transition of the early Universe. Both are topologically stable and both contain the massless chiral fermions (fermionic zero modes) localized in the cores of these linear defects. Here again the helium-3 vortices show a new phenomenon as compared with helium-4 vortices and Abrikosov vortices in conventional superconductors, and also with the proposed models for cosmic strings: in the ^{3}He-B quantized vortices the singularity of the phase Φ on the vortex axis is spontaneously dissolved.

This is the so-called flaring out of singularities from the vortex axis into extra dimensions: from the real space into extended (real and momentum) space. The singularities in the real space are thus transformed to the singularities in momentum space: point zeroes in the quasiparticle energy gap, or Fermi points.

The Fermi points are the counterpart of the Fermi surface – the surface in the momentum space of the normal Fermi liquid, where the quasiparticle energy spectrum crosses zero. Both Fermi points and Fermi surfaces, but not the Fermi lines, prove to be topologically stable objects in momentum space, i.e., they do not disappear under external perturbations of the ground state of condensed matter.

The topology of the Fermi points and Fermi surfaces, as part of a more general momentum-space topology of condensed matter, represents another important concept both for condensed matter and particle physics: internal topology of the vacuum state. The ground states may have the same symmetry, but differ by the value of the internal topological invariant, which results in different physical properties of the system. The nontrivial internal topology leads to such phenomena as quantization of different physical parameters and the fractional charge, spin and statistics of the particle-like objects.

In condensed matter it became important since the discovery of another amazing quantum object: a two-dimensional electron system displaying Integer and Fractional Quantum Hall Effect (IQHE and FQHE). In this system the kinetic parameter, the Hall conductivity σ_{xy}, is quantized in the fundamental units , the Planck constant h and electron charge e: $\sigma_{xy} = \frac{p}{q}\frac{e^2}{h}$ with integers p and q.

The superfluid phases of helium-3, with their unprecedented rich order parameter and nontrivial internal topology, also display the quantization of several physical parameters. The integer charges are expressed through the topological invariants in momentum space. The quantization is realized both in bulk liquids and in the quasi two-dimensional quantum object, the superfluid helium-3 film, where the quantum statistics of the particle-like soliton changes in stepwise manner from Fermi to Bose while the film thickness continuously increases. The transition from one quantum statistics to another occurs as the Lifshitz zero-temperature phase transition, when the thickness crosses some critical value, at which some integer topological invariant of ground state abruptly changes. The same topological invariant is responsible for the quantization of the response of the spin-current to the external perturbation, in the same manner as the electric current exhibits quantization in QHE.

For the investigation of such a complicated system as superfluid helium-3 which, with its multicomponent order parameter and the number of elementary particles and with the hierarchy of interactions and scales, may be compared to the Universe, the most useful guide is provided by symmetry and topology analyses. The main properties of a given superfluid phase are governed by the internal symmetry and internal topology of its vacuum state. The symmetry of the vacuum defines the long-wave properties of condensed matter, the quantum numbers of the elementary excitations and their mutual interaction, the possible topologically stable objects, etc., while the internal topology is responsible for the more intricate phenomena, including the quantization of physical parameters.

2
Broken Symmetry in Superfluid Phases of ³He

The phase transitions of liquid ³He into superfluid states, discovered by Osheroff, Richardson, and Lee (1972), are accompanied by a spontaneous breaking of symmetry, like any other phase transition of condensed matter into an ordered state.

2.1. Symmetry G of the Normal State of Liquid ³He

Above the critical phase-transition temperature T_c, liquid ³He is an isotropic, uniform, nonsuperfluid, nonchiral and nonmagnetic liquid, i.e., it has all the symmetries allowed in condensed matter. These symmetries form the symmetry group G of physical laws in condensed matter, which contains the following subgroups: 1) the group of translations t; 2) the group of solid rotations of coordinate space, SO_3; 3) the symmetry P under space parity transformation; 4) time inversion symmetry T; 5) the group $U(1)$ of gauge transformations representing gauge invariance.

The first three groups form the Euclidean group, the spontaneous breaking of this group leads to formation of such ordered condensed matter as crystals and liquid crystals, which have preferred positions of the atoms and/or preferred directions, the anisotropy axes. Time inversion symmetry T is broken in ordered magnets: antiferro- and ferromagnets, where the time inversion transformation does not leave the system invariant but reorients the ordered magnetic moments. Space inversion symmetry P is broken in ferroelectrics, where the spontaneous dielectric polarization arises, which changes sign under P transformation. And finally the breaking of the gauge symmetry group $U(1)$ gives rise to superfluidity and superconductivity, where the phase Φ of the complex order parameter ψ changes under the gauge transformation.

In some special cases, when the coupling between different subsystems of the condensed matter is small, additional elements of symmetry appear. For example, in some magnetic materials the spin-orbital coupling is small enough, as a result the separate spin rotations of the group $SO_3^{(S)}$ may be considered as symmetry operations, independent of the pure orbital rotation group $SO_3^{(L)}$. This is just the situation in liquid ^{3}He where the magnetic dipole interaction between the nuclear spins is negligibly small in comparison with the energies characterizing the superfluid transition.

All these symmetries except the translational one are simultaneously broken in liquid ^{3}He below T_c, where there spontaneously appears a Bose condensate of Cooper pairs which display internal degrees of freedom for both spin and orbital motion. This results in the unique and varied behaviour of the superfluid phases of ^{3}He which combine the properties of liquid crystals, ferro-, and antiferromagnets, "spin liquid crystals", and superfluids.

2.2. *Landau Theory of Superfluid Transition in* ^{3}He

According to the Landau theory of the second order phase transitions, the order parameter describing the spontaneous symmetry breaking in condensed matter is defined near T_c by one of the irreducible representations of the total group G: contributions from all other representations are relatively smaller in the vicinity of the transition in proportion to the small parameter $1 - T/T_c \ll 1$.

The order parameter in the pair correlated systems, such as ^{3}He and isotropic superconductors, transforms according to one of the irreducible representations described by the corresponding quantum numbers S and L of their Cooper pairs. This is the $(2S + 1) \times (2L + 1)$ dimensional representation of the group $SO_3^{(L)} \times SO_3^{(S)}$. For the isotropic superconductors this is the trivial one-dimensional representation with $S = L = 0$. The relevant representation for the order parameter in superfluid ^{3}He has quantum numbers $S = 1$ and $L = 1$, i.e., it is the $3 \times 3 = 9$-dimensional representation of the group $SO_3^{(L)} \times SO_3^{(S)}$.

The order parameter is thus defined by nine complex values. These are the pair amplitudes a_{M^S, M^L} of the eigenstates Ψ_{M^S, M^L} for the Cooper pair spin and orbital momentum projections with the eigenvalues $M^S = -1, 0, +1$ and $M^L = -1, 0, +1$ in the wave function of Cooper pair,

$$\Psi = \sum_{M^S, M^L} a_{M^S, M^L} \Psi_{M^S, M^L} \; .$$

At $T > T_c$ all the pair amplitudes are zero in equilibrium, since no Bose condensate of the Cooper pairs exists in the normal liquid. Below T_c the instability develops towards a simultaneous increase of all $(2S+1) \times (2L+1)$ complex amplitudes a_{M^S, M^L} of a given representation, according to the quadratic term in the Ginzburg-Landau free energy functional

$$\sim (T - T_c) \sum_{M^S, M^L} |a_{M^S, M^L}|^2 \; ,$$

which becomes negative at $T < T_c$.

In the case of the 3×3 dimensional representation, instead of this set of amplitudes a_{M^S, M^L}, it is more convenient to utilize the vector representation for each of the two unit angular momenta, $S = 1$ and $L = 1$, by organizing these nine complex amplitudes in tensorial form, which have more transparent transformation properties. In the vector representation the amplitudes a_-, a_0, a_+ of the corresponding projections $M = -1, 0, +1$ of unit momentum may be rearranged to form the components of the vector a_i: $a_\pm = (a_x + i a_y)/\sqrt{2}$ and $a_z = a_0$. As a result, we come to the commonly used 3×3 complex matrix order parameter in superfluid ^{3}He:

$$A_{\alpha i} = \sum_{M^S, M^L} a_{M^S, M^L} \lambda_\alpha^{M_S} \lambda_i^{M_L} \; ,$$

where $\lambda_\alpha^0 = \hat{z}_\alpha$, $\lambda_\alpha^\pm = (\hat{x}_\alpha \pm i \hat{y}_\alpha)/\sqrt{2}$. This matrix transforms as a vector under a spin rotation for a given orbital index (i), and as a vector under an orbital rotation for a given spin index (α). So the quadratic term in the G-L functional is proportional to $(T - T_c)\, A^*_{\alpha i} A_{\alpha i}$. More formal definition of the order parameter $A_{\alpha i}$ is given in Sec. 5.3.

The growth of the amplitudes $A_{\alpha i}$ is limited by the fourth-order terms in the Ginzburg-Landau free-energy functional which stabilize the amplitudes in the proportion, corresponding to some superfluid phase. The Ginzburg-Landau free-energy functional must be invariant under the total symmetry group G of the physical laws. This invariance essentially restricts the number of the fourth-order terms in the bulk energy and also the number of the gradient terms. The bulk condensation energy term in the G-L free-energy functional, $F^{\text{G-L}}_{\text{bulk}}[A_{\alpha i}]$, in superfluid ^{3}He contains one second-order term, discussed above, and five fourth-order terms. In each of them the Greek spin indices should not be mixed with the Latin orbital indices to provide the invariance under separate spin $SO^{(S)}_3$ and orbital $SO^{(L)}_3$ rotations; also each term should contain an equal number of $A^*_{\alpha i}$ and $A_{\alpha i}$ to satisfy gauge invariance. As a result of these requirements $F^{\text{G-L}}_{\text{bulk}}[A_{\alpha i}]$ is given by

$$F^{\text{G-L}}_{\text{bulk}} = -\alpha A^*_{\alpha i} A_{\alpha i} + \beta_1 A^*_{\alpha i} A^*_{\alpha i} A_{\beta j} A_{\beta j} + \beta_2 A^*_{\alpha i} A_{\alpha i} A^*_{\beta j} A_{\beta j} + \beta_3 A^*_{\alpha i} A^*_{\beta i} A_{\alpha j} A_{\beta j}$$

$$+ \beta_4 A^*_{\alpha i} A_{\beta i} A^*_{\beta j} A_{\alpha j} + \beta_5 A^*_{\alpha i} A_{\beta i} A_{\beta j} A^*_{\alpha j}\,, \qquad (2.1)$$

where α changes sign at T_c, $\alpha = \alpha_0(1 - T/T_c)$, while α_0 and the β's are functions of pressure only and depend on the details of the microscopic interaction of the ^{3}He atoms. The superfluid phases below T_c are found now from the minimization of this functional.

2.3. Classes of Superfluids

The equilibrium order parameter $A^0_{\alpha i}$ is found as the solution of the Ginzburg-Landau equations $\delta F^{\text{G-L}}_{\text{bulk}}/\delta A_{\alpha i} = 0$. The solution, which realizes the absolute minimum of $F^{\text{G-L}}_{\text{bulk}}$, is $A_{\alpha i} = 0$ above T_c and this corresponds to the normal liquid. Below T_c, i.e., in the superfluid state, there are many solutions of the G-L equations with $A^0_{\alpha i} \neq 0$. Depending on the relations between the β parameters some of the solutions correspond to the local minima, while the others realize the saddle points of the G-L functional. The symmetry arguments are the best way for the classification of the solutions.

It is important that some of these solutions may be obtained from each other by one of the symmetry operations g of the group G. In this case we say that they correspond to the same superfluid phase but to different degenerate physical states of this phase. Indeed, if some $A^0_{\alpha i} \neq 0$ realizes the minimum of the functional F^{G-L}_{bulk}, then, according to the symmetry G of the functional, the state $(gA^0)_{\alpha i}$, obtained from $A^0_{\alpha i}$ by action of any element g of the group G, has the same energy and therefore is also an equilibrium state.

Also such pairs of solutions exist which in no way can be transformed into each other, and we refer to them as belonging to different superfluid phases. So the solutions are divided into classes of states: each class describes the corresponding superfluid phase of ^3He and this class contains all the degenerate states of this phase. The configurational space of all the degenerate states within a given class may be referred to as a manifold of degenerate equilibrium states or a manifold of internal states. For example the manifold of internal states for superfluid ^4He is the circumference S^1, at which the phase Φ of the complex order parameter ψ varies.

Each class is characterized by its symmetry. For a given equilibrium state (or vacuum state) of the ordered phase there exist some elements of the group G which do not change the state, i.e., leave the order parameter invariant. These elements form the group H of the residual symmetry of the equilibrium state, which is a subgroup of G. All the most important physical properties of a given superfluid phase (or more generally of a given ordered condensed matter), which distinguish it from the other systems, are mostly determined by the group H of its equilibrium state. Since there are many subgroups H of the group G, one may conclude that there are many types of vacuum states, which thus correspond to different ordered phases with different internal symmetries of vacuum state. Hence the symmetry classification of the possible superfluid phases is just reduced to enumeration of all those groups $H \in G$ which do not contain $U(1)$ as a subgroup.

2.4. *Equilibrium Order Parameters for A- and B- Phases of* ^3He

Which symmetry class H, i.e., what kind of superfluids, has absolute minimum of the free energy and thus is realized under given external conditions depends on the details of the Ginzburg-Landau functional, in our case, on the relations between the coefficients β's of the fourth-order invariants in Eq. (2.1). One may regulate the β's by changing the external conditions, e.g. by applying different pressures, and thus stimulate the phase transitions between different superfluid phases. Microscopic theory still does not give precise values for the β's. However experimentally we know that there are two intervals of pressures where two different superfluid phases, A and B, take place, and from all the experimental data one may identify them with definite extremal points of the Ginzburg-Landau functional.

Below the so called policritical point (PCP), $0 \leq P < P_{PCP}$ (Fig. 1.1b), the relations between the β parameters are such that the most energetically favourable equilibrium order parameter corresponds to the following solution of the Ginzburg-Landau equations $\delta F_{bulk}^{G-L}/\delta A_{\alpha i} = 0$:

$$A_{\alpha i}^0 = \Delta_B(T, P)\delta_{\alpha i} . \tag{2.2}$$

Here

$$\Delta_B^2(T, P) = \frac{\alpha}{2(\beta_{345} + 3\beta_{12})} , \tag{2.3}$$

and the symbol $\beta_{i...j}$ denotes the sum $\beta_i + \cdots + \beta_j$ of the corresponding β's. In terms of the amplitudes a_{M^S, M^L} of the spin and orbital momentum projections this is a mixture of three equally weighted interpenetrating superfluids, each with zero projection of the total spin, $M^J = M^L + M^S = 0$:

$$a_{+-} = a_{-+} = a_{00} = \Delta_B(T, P) , \tag{2.4}$$

which corresponds to the isotropic state with $J = 0$. This is the isotropic B-phase of superfluid ^{3}He.

For pressures above policritical point the A-phase takes place which corresponds to the absolute minimum of Eq. (2.1) with the following equilibrium order parameter:

$$A_{\alpha i}^0 = \Delta_A(T, P)\hat{z}_\alpha(\hat{x}_i + i\hat{y}_i) , \tag{2.5}$$

where $\hat{z}$, $\hat{x}$ and $\hat{y}$ are axes of the Cartesian coordinate frame, and

$$\Delta_A^2(T, P) = \frac{\alpha}{4\beta_{245}} . \tag{2.6}$$

In terms of the amplitudes a_{M^S,M^L} only one of the amplitudes in the A phase is nonzero:

$$a_{0+} = \sqrt{2}\Delta_A(T,P) , \qquad (2.7)$$

i.e., this is the state with $M^S = 0$, $M^L = 1$, where the quantization axes for spin and orbital momentum were chosen both along the same axis $\hat{z}$.

2.5. Degeneracy of Equilibrium States

These two phases differ by their symmetry and, as a consequence, by the manifold of degenerate equilibrium states. Let us first consider the manifolds of the internal states for these two phases. To obtain all the degenerate states of a given superfluid phase one should apply continuous symmetry operations **g** from the group $U(1) \times SO_3^{(L)} \times SO_3^{(S)}$ to some chosen initial equilibrium state of the phase: for B and A phases these initial states are in Eqs. (2.2) and (2.5) correspondingly. If R_{ij}^L and $R_{\alpha\beta}^S$ are matrices of solid rotations $SO_3^{(L)}$ and $SO_3^{(S)}$ in orbital and spin spaces and Φ is the parameter of the global gauge transformation $U(1)$, then all the degenerate equilibrium states are obtained from some initial one in the following way:

$$(\mathbf{g}A^0)_{\alpha i} = \exp(i\Phi)R_{ij}^L R_{\alpha\beta}^S A_{\beta j}^0 . \qquad (2.8)$$

(To be more exact, the gauge transformation with the parameter α multiplies the order parameter by the phase factor $e^{2i\alpha}$ since the Cooper pair contains two particles. We shall take this into account further when it is necessary.)

Applying this to the initial states in Eqs.(2.2),(2.5) one obtains the general form of the equilibrium order parameter in the B- and A- phases:

$$A_{\alpha i}^0(\text{B-phase}) = \Delta_B(T,P)\exp(i\Phi)R_{\alpha i} , \qquad (2.9)$$

$$A_{\alpha i}^0(\text{A-phase}) = \Delta_A(T,P)\hat{d}_\alpha(\hat{e}_i^{(1)} + i\hat{e}_i^{(2)}) , \qquad (2.10)$$

which describes all the possible equilibrium states in the two phases.

2.6. Manifold of Degenerate States in ^{3}He-B

In superfluid ^{4}He and conventional superconductors, the only degeneracy parameter is the phase Φ of the Bose condensate. The manifold of internal states is a circumference S^1 (or $U(1)$), which is the same at which the phase Φ changes. In both superfluid phases of ^{3}He the manifold of degenerate states is essentially more complicated: there are several parameters which enumerate the degenerate states.

In the superfluid ^{3}He-B in addition to the circumference S^1 of the phase Φ the degenerate states are also labelled by the orthogonal real matrix $R_{\alpha i}$, which is obtained from the initial unit matrix $\delta_{\alpha i}$ in Eq. (2.2) by orbital or spin rotations, or by a combination of rotations:

$$R_{\alpha i} = R^L_{ij} R^S_{\alpha\beta} \delta_{\beta j} , \quad \text{or} \quad \mathbf{R} = \mathbf{R}^S (\mathbf{R}^L)^{-1} . \tag{2.11}$$

As distinct from the initial equilibrium state in Eq. (2.2), the equilibrium state, described by matrix $\mathbf{R}$, is no more a state with $J = 0$. According to Eq. (2.11), the matrix $\mathbf{R}$ defines the *relative* rotation of spin and orbital frames, which should be applied to the initial state with $J = 0$ to obtain the given state.

It is of great importance for further results that this superfluid system is sensitive only to the relative rotations of spin and orbital frames. This is the phenomenon of *broken relative symmetry*, which will be extensively discussed below for both the superfluid phases, since it results in the most intriguing properties of these liquids.

Thus the manifold of the equilibrium degenerate states in the superfluid ^{3}He-B (we denote it as R_B) has dimension 4: one-dimensional circumference S^1, where the phase Φ varies, times the three-dimensional space SO_3^{relative} of relative spin-to-orbit rotations produced by matrix $\mathbf{R}$:

$$R_B = S^1 \times SO_3^{\text{relative}} . \tag{2.12}$$

2.7. Magnetic Anisotropy in ^{3}He-A

In the superfluid ^{3}He-A the degenerate states in Eq. (2.10) are labelled by two different degeneracy parameters which describe separately the spin and orbital degrees of freedom of equilibrium states.

The spin subsystem is described by a real unit vector $\hat{d}$ which is obtained from spin rotations of the initial axis $\hat{z}$ of spin quantization: $\hat{d}_\alpha = R^S_{\alpha\beta} \hat{z}_\beta$.

The appearance of the vector $\hat{d}$ means breaking of the $SO_3^{(S)}$ symmetry, because $\hat{d}$ gives the specific direction in the liquid which initially, above T_c, had absolutely isotropic magnetic properties. Now the vector $\hat{d}$ indicates the direction of the spontaneous magnetic anisotropy. This means that the magnetic susceptibility in the A-phase is a uniaxial tensor as in collinear antiferromagnets:

$$\chi_{\alpha\beta} = \chi_\parallel \hat{d}_\alpha \hat{d}_\beta + \chi_\perp (\delta_{\alpha\beta} - \hat{d}_\alpha \hat{d}_\beta) \ . \tag{2.13}$$

The direction of $\hat{d}$ is arbitrary, which is the essence of the degeneracy in an ordered system with broken continuous symmetry; the change of its orientation costs no energy. There are several notations for such variables (degeneracy parameter, soft variable, Goldstone mode, hydrodynamical variable) which we shall use. Since there are no restoring forces, the Goldstone variable $\hat{d}$ may be easily oriented by external fields. For example, according to the magnetic anisotropy in Eq. (2.13) there is a term in the magnetic energy

$$-\frac{1}{2}\chi_{\alpha\beta}H_\alpha H_\beta = -\frac{1}{2}\chi_\perp H^2 + \frac{1}{2}(\chi_\perp - \chi_\parallel)(\hat{d} \cdot \vec{H})^2 \ , \tag{2.14}$$

i.e. the second term on the rhs of Eq. (2.14), which is responsible for the orientational effect of the magnetic field on the $\hat{d}$ vector. The magnetic anisotropy is large in the A-phase (of order unity) as in conventional antiferromagnets. At low temperatures $\chi_\parallel \to 0$ (Leggett, 1975). Since $\chi_\perp > \chi_\parallel$ the spin axis $\hat{d}$ is orientated at equilibrium in a plane which is transverse to a magnetic field – provided that the field is large enough to neglect other orienting effects, e.g. tiny spin-orbital coupling.

2.8. *Liquid-crystal Anisotropy in ³He-A*

The orbital part of the order parameter in Eq. (2.10) is a complex vector $\hat{e}^{(1)} + i\hat{e}^{(2)}$, obtained from the initial complex vector $\hat{x} + i\hat{y}$ both by orbital rotations and gauge transformation:

$$\hat{e}_i^{(1)} + i\hat{e}_i^{(2)} = \exp(i\Phi)R_{ij}^L(\hat{x}_j + i\hat{y}_j) \ . \tag{2.15}$$

As follows from Eq. (2.15) $\hat{e}^{(1)}$ and $\hat{e}^{(2)}$ are unit vectors and orthogonal to each other.

The vector $\hat{l}_i = R^L_{ij}\hat{z}_j$, which is obtained by the orbital rotation of the orbital momentum quantization axis, defines the direction of the Cooper pair orbital angular momentum in the state given by Eq. (2.10). It is expressed in terms of these two vectors: $\hat{l} = \hat{e}^{(1)} \times \hat{e}^{(2)}$. Since the orbital vector $\hat{l}$ changes sign under time inversion (from $\mathbf{T}(\hat{e}^{(1)} + i\hat{e}^{(2)}) = \hat{e}^{(1)} - i\hat{e}^{(2)}$ it follows that $\mathbf{T}\hat{l} = -\hat{l}$), it produces the orbital ferromagnetism along $\hat{l}$.

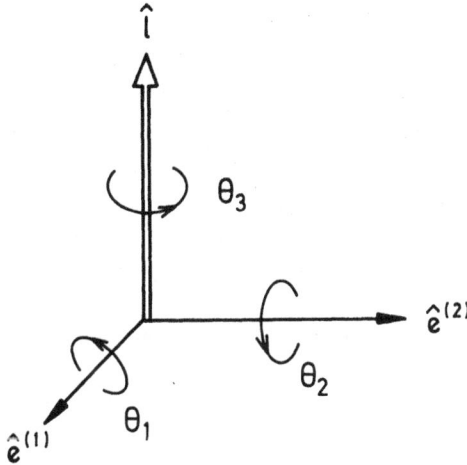

Fig. 2.1. The orbital part of the degeneracy parameter of the A-phase defines the local coordinate frame of the A-phase vacuum. This is the origin of the gravitational field in this phase. The rotation of the *dreibein* by the angle θ_3 about the anisotropy axis $\hat{l}$ has the same result as a gauge transformation.

Three orthonormal vectors $\hat{l}$, $\hat{e}^{(1)}$ and $\hat{e}^{(2)}$ form the *dreibein* − a local physical coordinate frame of the A-phase vacuum. The existence of the distinguished coordinate frame is the manifestation of the breaking of the $SO_3^{(L)}$ symmetry. Among the three solid rotations of the *dreibein* the rotation θ_3 about axis $\hat{l}$ is different from the other two rotations about $\hat{e}^{(1)}$ and $\hat{e}^{(2)}$. Such rotation can completely compensate the gauge transformation of the *dreibein* (see Sec. 2.11 below). So, again, now in the A-phase, we come across an example of the broken relative symmetry. This broken symmetry is related to the gauge transformation: only the gauge transformation counted from the orbital rotations about axis $\hat{l}$, i.e., relative gauge orbital transformation, changes the degenerate state. This may be called broken relative gauge-orbital symmetry.

The first consequence of this breaking of the relative gauge-orbital symmetry is as follows: all those physical variables in the A-phase which are gauge invariant should be invariant also under rotations about axis $\hat{l}$. So for these variables only one specific direction exists – the axis $\hat{l}$: they feel the breaking of the $SO_3^{(L)}$ invariance as the appearance of the anisotropy axis $\hat{l}$. Therefore the vector $\hat{l}$ defines the spontaneous orbital liquid-crystal-like anisotropy in ^{3}He-A; all the gauge invariant physical quantities related to the orbital subsystem of ^{3}He-A have uniaxial anisotropy: the viscosity, the normal fluid and superfluid densities, heat conductivity, dielectric constant, etc. For example the kinetic energy of the superfluid motion is anisotropic in the A-phase:

$$\frac{1}{2}(\rho_s)_{ij}(\vec{v}_s)_i(\vec{v}_s)_j = \frac{1}{2}\rho_s^\perp \vec{v}_s^2 + \frac{1}{2}(\rho_s^\| - \rho_s^\perp)(\hat{l} \cdot \vec{v}_s)^2 \, , \qquad (2.16)$$

where $\rho_s^\|$ and $\rho_s^\perp$ denote the longitudinal and transverse components of the superfluid density tensor:

$$(\rho_s)_{ij} = \rho_s^\| \hat{l}_i \hat{l}_j + \rho_s^\perp (\delta_{ij} - \hat{l}_i \hat{l}_j) \, . \qquad (2.17)$$

The direction of $\hat{l}$ is arbitrary in the same way as is that of $\hat{d}$, also $\hat{l}$ may be easily oriented by external fields. For example, according to the anisotropy of superflow in Eq. (2.16) the superfluid velocity produces an orienting effect on the $\hat{l}$ vector. Since $\rho_s^\| < \rho_s^\perp$ the $\hat{l}$ vector is aligned with the superfluid velocity. Another orientating effect comes from the tiny spin-orbital coupling between $\hat{d}$ and $\hat{l}$:

$$F_{\rm so} = -\frac{1}{2}g_{\rm so}(\hat{d} \cdot \hat{l})^2 \, . \qquad (2.18)$$

Then, as distinct from the spin vector $\hat{d}$, the orbital $\hat{l}$ vector interacts with the vessel walls, resulting in the boundary conditions for $\hat{l}$: it should be oriented along the normal to the boundary.

2.9. Manifold of Degenerate States in ^{3}He-A

Thus in the superfluid ^{3}He-A the manifold of the equilibrium degenerate states has dimension 5: two-dimensional sphere S^2, where the unit vector $\hat{d}$ varies, times the three-dimensional space SO_3 of solid rotations of the orthonormal dreibein $\hat{l}$, $\hat{e}^{(1)}$ and $\hat{e}^{(2)}$. However this is still not the whole story; it is not correct to write the manifold R_A of internal states in the superfluid ^{3}He-A as a simple product of the above spaces, since there will be double counting. Each physical state with given $\hat{d}$ and $\hat{e}^{(1)} + i\hat{e}^{(2)}$ may also be represented by another set of the degeneracy parameters, namely by the opposite vectors $-\hat{d}$ and $-(\hat{e}^{(1)} + i\hat{e}^{(2)})$, since the order parameter in Eq. (2.10) does not change if we reverse the signs of both degenerate parameters. Therefore the product $S^2 \times SO_3$ should be factorized by the group Z_2 of two points:

$$R_A = (S^2 \times SO_3)/Z^2 \, , \qquad (2.19)$$

to produce the space of the degenerate states.

2.10. Residual Symmetry H of ^{3}He-B

Equations (2.12) and (2.19) for the space of degenerate states in A and B phases may be directly obtained from the residual symmetry H of these phases, which we discuss here. The group H consists of those elements of the symmetry group G, which leave the order parameter unchanged. In ^{3}He-B, according to Eq. (2.11) the degeneracy parameter matrix $\mathbf{R}$ is unmoved if the spin rotation $\mathbf{R}^S$ is accompanied by an equal rotation $\mathbf{R}^L$ of the orbital space. This just manifests that only the relative spin-orbit symmetry is broken. Therefore the subgroup H of the symmetry group $G = U(1) \times SO_3^{(L)} \times SO_3^{(S)}$ is just the group $SO_3^{(J)}$ of the combined rotations. The conservation of the combination of two or more symmetries with the breaking of the separate symmetries is the essence of the broken relative symmetry.

The manifold of degenerate states may thus be found from the groups G and H using the fact that the elements of the subgroup H leave the given equilibrium state invariant and, therefore, do not participate in producing new degenerate states from the initial one. So in the process of creating the new degenerate states $\mathbf{g}A^0$ by the action of the elements of G, one may reduce the whole subgroup H into the identity element. Therefore the space of degenerate equilibrium states is just the factor space G/H. In the B-phase case the factorization of the product of rotation groups $SO_3^{(L)} \times SO_3^{(S)}$ by the group of combined rotations just gives the group of relative rotations, which enumerates the rotational degrees of freedom of the B-phase:

$$R_B = G/H = (U(1) \times SO_3^{(L)} \times SO_3^{(S)})/SO_3^{(J)} = S^1 \times SO_3^{\text{relative}} . \quad (2.20)$$

2.11. Residual Symmetry H of ^{3}He-A

In the A-phase there are several different constituents of the group H which do not change the order parameter in Eq. (2.10):

i) The spin rotations $SO_2^{(S)}$ about the axis $\hat{d}$.

ii) According to Eq. (2.15) there is a very important combined continuous symmetry $U(1)^{\text{combined}}$, which manifests the broken relative gauge-orbital symmetry: the A-phase state is invariant under gauge transformation with the parameter Φ from $U(1)$ group if it is simultaneously accompanied by orbital rotation of the dreibein $\hat{e}^{(1)}$, $\hat{e}^{(2)}$ and $\hat{l}$ about axis $\hat{l}$ by the same angle $\theta_3 = \Phi$. Under orbital rotation about $\hat{l}$ the orbital degeneracy parameter is multiplied by the phase factor:

$$\hat{e}^{(1)} + i\hat{e}^{(2)} \rightarrow e^{-i\theta_3}(\hat{e}^{(1)} + i\hat{e}^{(2)}) ,$$

in the same manner as under a gauge transformation. So, if $U(1)$ is combined with this rotation by $\theta_3 = \Phi$, the order parameter does not change.

This symmetry is responsible for many exotic properties of the A-phase: it is responsible for the nodes in the gap in the energy spectrum of fermions; it gives rise to the chirality of fermions and to the chiral anomaly; it leads to the non-potential character of the A-phase superflow, and results in the continuous vortex texture in the rotating vessel.

iii) The discrete combined symmetry Z_2^{combined}, which was already discussed in Sec. 2.7; the spin rotation of $\hat{d}$ by angle π about a perpendicular axis changes the sign of $\hat{d}$, but this may be compensated by a gauge transformation with $\Phi = \pi$, which changes the sign of $\hat{e}^{(1)} + i\hat{e}^{(2)}$ according to Eq. (2.15) and thus leaves invariant the whole order parameter in Eq. (2.10). This combined symmetry, as will be shown below in Sec. 7.3, leads to the existence of quite unusual half-quantum vortices.

Two more discrete combined symmetries, which are not important here, will be discussed later in Sec. 4 in connection with the classification of quasi-particles.

Now taking into account that $SO_3^{(S)}/SO_2^{(S)} = S^2$ one obtains for the space of degenerate equilibrium states of the A-phase the same Eq. (2.19):

$$R_A = G/H = (U(1) \times SO_3^{(L)} \times SO_3^{(S)})/(SO_2^{(S)} \times U(1)^{\text{combined}} \times Z_2^{\text{combined}})$$

$$= (S^2 \times SO_3^{\text{relative}})/Z_2 . \tag{2.21}$$

Here SO_3^{relative} means that only such rotations about the axis $\hat{l}$ lead to new degenerate states which are relative towards gauge transformations. So in both phases there is the phenomenon of broken relative symmetry with con-servation of the combined one: in the B-phase it is a continuous symmetry of the relative spin-to-orbit rotations which is broken while the system is invariant under combined rotations; in the A-phase the same phenomenon occurs with gauge and orbital rotations, and with discrete gauge and spin rotations.

Here we found the symmetry of given superfluid phases from the explicit structure of the equilibrium order parameter obtained from the solution of the Ginzburg-Landau equations which is valid only near T_c. What occurs far below T_c where an admixture of the other representations becomes im-portant and the order parameter is not described by a 3×3 matrix? In the whole region of existence of a given phase the symmetry H is the same by the definition of the phase, since the symmetry of state can change only at a phase transition line between different phases. So all the properties of the superfluid phase, which are related to the symmetry H, are conserved also far from T_c. This includes also the degeneracy parameters, which are the same in the whole region of the given phase.

So the general procedure which we used within the Landau scheme is as follows: 1) one chooses the relevant irreducible representation; 2) in the vicinity of transition one writes the G-L functional for the order parameter which transforms according to this irreducible representation; 3) one finds the solution of G-L equations, which corresponds to the absolute minimum of the functional; 4) one finds the symmetry H of this solution and the degeneracy parameters, describing the equilibrium states; the results of this stage are valid not only in the vicinity of T_c but at all temperatures; 5) one writes the London energy functional for the degeneracy parameters (see Sec. 3).

2.12. *Combined Spin-orbital Symmetry and Relative Spin-orbital Anisotropy in ³He-B*

The phenomenon of broken relative symmetry in the isotropic ³He-B leads to a unique situation. Both the orbital and spin properties of this superfluid are isotropic, there is no specific direction either in spin or in orbital space. Nevertheless, there occurs a kind of relative anisotropy, represented by the degeneracy parameter matrix $R_{\alpha i}$. This relative anisotropy of the magnetic and orbital properties in the isotropic liquid is manifested through the following effect: if any anisotropy axis $\hat{n}^{(L)}$ of the orbital properties is induced by external conditions, there simultaneously appears a magnetic anisotropy axis. However, the direction $\hat{n}^{(S)}$ of the magnetic axis is not given by $\hat{n}^{(L)}$, but is obtained by the relative rotation of the orbital axis: $\hat{n}_\alpha^{(S)} = R_{\alpha i}\hat{n}_i^{(L)}$ is the anisotropy axis for the magnetic properties. For example, the container wall produces an orbital anisotropy of the B-phase in the surface layer with the natural anisotropy axis along the normal $\hat{\nu}$ to the wall, $\vec{n}^{(L)} = \hat{\nu}$. This automatically results in a magnetic anisotropy axis $\hat{n}_\alpha^{(S)} = R_{\alpha i}\hat{\nu}_i$. It means that near the wall the magnetic susceptibility is a uniaxial tensor like in the A-phase in Eq. (2.13), where the A-phase magnetic anisotropy axis $\hat{d}$ should be substituted for the B-phase magnetic anisotropy axis $\hat{n}^{(S)}$.

3
Textures and Supercurrents in Superfluid Phases of ^{3}He

3.1. *Textures, Gradient Energy and Rigidity*

The broken symmetry in the ordered system leads to the phenomenon of inhomogeneous states with spatially distributed order parameter, which are usually referred to as textures. The texture corresponds to inhomogeneous vacuum in quantum field theory. Here we discuss first the textures of the degeneracy parameters, i.e., of soft Goldstone variables. It means that everywhere in the texture the state is locally described by the equilibrium order parameter of the given superfluid phase, and what does change in this texture is the orientation of the order parameter, described by the degeneracy parameters. So the system nowhere leaves the manifold of the internal states of the given superfluid phase. As will be seen below, such a situation, known as the London limit, occurs if the length scale at which the degeneracy parameters change in texture is large compared with some characteristic length (the coherence length ξ).

The textures may arise as a result of competition between different orienting factors, such as external fields, spin-orbital coupling and boundary conditions, or may exist as metastable state, like quantized vortex or domain wall, whence stability is prescribed by conservation of some topological invariants. In some cases just the nonuniform texture realizes the state with absolute minimum of the free energy functional.

For example the equilibrium state of the A-phase in a vessel of spherical form cannot be uniform because of competing (frustrating) boundary
conditions for the $\hat{l}$ vector, which should be normal to the wall everywhere
in the vessel. Three possible inhomogeneous ground states are shown in
Fig. 3.1. Moreover the shape of the vessel is topologically incompatible with
the continuous distribution of the $\hat{l}$ texture: the spherical shape of the vessel, combined with the normal boundary conditions on the surface of the
vessel, requires the existence of a singular point in the vector $\hat{l}$ field. This
point may either be on the surface (Fig. 3.1c), and this surface point defect
is called a boojum, or inside the vessel (Figs. 3.1a and 3.1b). As we shall see
later in Sec. 3.7 such point defect inside the vessel should be accompanied
by one or two tails in the form of the linear defects, the quantized vortices.
This resembles the Dirac magnetic monopole, especially if one compares the
distribution of the superfluid velocity around this monopole-like defect and
the electromagnetic field $\vec{A}$ around the magnetic monopole (Sec. 7.11).

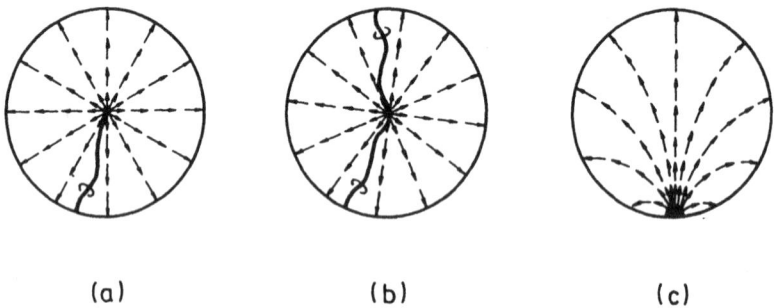

(a)   (b)   (c)

Fig. 3.1. The $\hat{l}$-textures in a spherical vessel. The radial distribution of the $\hat{l}$
vector (a) corresponds to the Dirac magnetic monopole, it gives rise to a string –
the doubly quantized vortex, with one nd terminating on the monopole and the
other end on the surface of the container. The doubly quantized vortex may split
into two singly quantized vortices (b) or may contract to the point (c), in the latter
case the monopole transforms to the point defect on the surface, the boojum.

However if one can discard the boundary conditions, a nonuniform texture usually has higher energy than the uniform ground state due to the

energy of inhomogeneity, referred to usually as the gradient energy. In the absence of orientational effects the free energy of the ordered system does not depend on the orientation of the degeneracy parameters, but should depend on their gradients.

For example the free energy of the superfluid ^{4}He cannot depend on the phase Φ of the order parameter, since the energy should be invariant under the global gauge transformation $\Phi \rightarrow \Phi + \alpha$. However this symmetry does not prohibit the dependence of energy on the gradient of Φ, since $\vec{\nabla}\Phi$ is invariant under the global gauge transformation. To the lowest (second) order in gradients this dependence may be written as:

$$F_{\text{grad}}^{\text{London}} = \frac{1}{2}K(\vec{\nabla}\Phi)^2 \ , \tag{3.1}$$

where the notation *London* in the superscript means that the London limit is considered, when only the energy dependence on the degeneracy parameters is taken into account, while the modulus of ψ is supposed to be in equilibrium.

The coefficient K in the gradient energy is usually referred to as rigidity, or stiffness. This terminology comes from the crystals, where the degeneracy parameter is the vector $\vec{u}$ of the displacement of atoms of the crystal lattice, and the corresponding coefficients form the elastic tensor Λ_{ijkl} in the elastic energy of crystal (which is the London energy for the crystals):

$$F_{\text{grad}}^{\text{London}} = \frac{1}{2}\Lambda_{ijkl}u_{ij}u_{kl} \ . \tag{3.2}$$

They describe the rigidity of the crystal towards the strain u_{ik} , which is expressed in terms of the gradients of the degeneracy parameter $\vec{u}$, namely, $u_{ik} = \partial_i u_k + \partial_k u_i$.

In the general case of condensed matter with broken symmetry, the rigidity describes the response of the system towards the strain which is the gradient of the relevant inhomogeneous distortion: gradient of displacement in crystals, gradient of the gauge transformation in superfluids and superconductors, gradient of the solid angle, $\nabla_i \vec{\theta}$, of the orbital and spin rotation for liquid crystals and ordered magnet, respectively. Such response is absent in a disordered state.

The concept of rigidity is one of the most important features of ordered systems, and even more important than the existence of the order parameter. In some cases, e.g. in two-dimensional systems, the fluctuations suppress the long-range order, and as a result no order parameter exists in the system, while the rigidity is still retained and is responsible for most of the properties of the "ordered" phase. Such a state without the order parameter but with rigidity is known as a Berezinskii-Kosterlitz-Thouless state.

3.2. *Why Superfuids are Superfluid*

In superfluids the gradient of phase is related to the frictionless mass motion and the London energy is the kinetic energy of this coherent cooperative motion of the Bose condensate:

$$F_{\text{grad}}^{\text{London}} = \frac{1}{2}\rho_s \vec{v}_s^2 \quad , \quad \vec{v}_s = \frac{\hbar}{m_4}\vec{\nabla}\Phi \ .$$

The introduced vector $\vec{v}_s$ has the property of a velocity: it properly transforms under a Galilean transformation. Under the Galilean transformation to a frame moving with the velocity $\vec{w}$, the wave function changes as $\psi \rightarrow \psi \exp(im_4\vec{w}\cdot\vec{r}/\hbar)$, as a result the superfluid velocity transforms as $\vec{v}_s \rightarrow \vec{v}_s + \vec{w}$. The coefficient $\rho_s = K(m_4/\hbar)^2$ is thus some mass density which is the density of superfluid component of the liquid, that part of the liquid which moves without any friction.

The absence of friction comes from the fact that this mass flow results from the inhomogeneous texture in the Φ field, and we know that the order parameter inhomogeneity cannot be a source of the energy dissipation, since the nonuniform texture may occur even in the ground state of the ordered system, as we have seen for the example of the A-phase in the spherical vessel. So just this fact that the textures do not produce the dissipation is in the basis of the frictionless flow of superfluids, produced by the texture of the phase Φ of the Bose condensate: since the flow is produced by the equilibrium texture, this flow cannot dissipate.

One can easily imagine a metastable equilibrium state with the nonzero phase gradient $\vec{\nabla}\Phi$ of the order parameter, and therefore with nonzero mass current $\vec{j} = \rho_s\vec{v}_s$, in superfluid ^4He. This is the state of the superfluid ^4He

in a channel with topologically nontrivial connection, e.g., in a form of torus (Fig. 3.2).

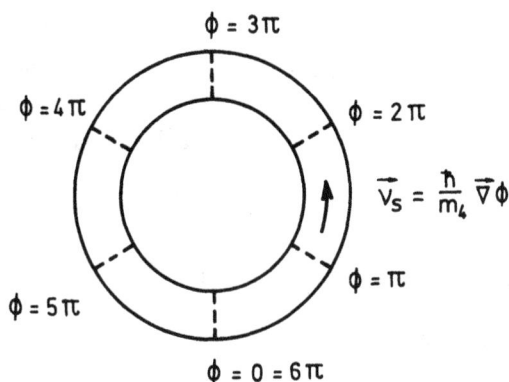

Fig. 3.2. The superflow with winding number $m_\Phi = 3$ in the closed channel.

If the phase Φ has $2\pi m_\Phi$ winding along a torus-shaped channel, where m_Φ should be an integer to provide the continuity of the order parameter in the channel, then one cannot decrease m_Φ continuously. To transfer a flow state with the given m_Φ to another state with, say, $m_\Phi \pm 1$, one should break somewhere in the channel the continuous distribution of the phase Φ. This violation of the coherent vacuum state costs energy, therefore energy barriers exist between states with different N. This means that the state with given $m_\Phi \neq 0$ (and therefore with the nonzero $\vec{\nabla}\Phi$) is highly stable: though this state corresponds to the local minimum of the energy which is higher than the absolute minimum of the homogeneous state, there is practically no chance to eliminate this state with the mass flow. So the stationary gradient of the phase, as well as the gradient of any other degeneracy parameter in condensed matter, cannot be a source of dissipation. The mass current $\vec{j} = \rho_s \vec{v}_s$ may thus circulate without any friction, and this is the essence of the phenomenon of superfluidity.

The dissipation of the superflow or other textures in condensed matter occurs only above some threshold when the gradients of the order parameter become large enough to overcome the energy barrier between the states with different m_Φ. In this case instability develops towards the time-dependent evolution of the order parameter at which unwinding of m_Φ takes place. This is the so-called phase slippage process.

The supercurrent transfers only some part of the liquid mass, the super-fluid component, while the other part of the liquid participates in conventional dissipative motion like the normal liquid above T_c. This part forms the normal component of the liquid with the density ρ_n. So the total current is:

$$\vec{j} = \rho_s \vec{v}_s + \rho_n \vec{v}_n ,$$

with $\rho_n + \rho_s = \rho$, the total mass density.

3.3. *Superfluidity and Response to a Transverse Gauge Field*

The supercurrent vanishes when the generalized rigidity ρ_s of the texture is absent, so to find out if there is a superfluidity in a given system one should calculate ρ_s. Here we discuss a general formalism of the response of the system to the compensating gauge fields, which is commonly used to find the rigidity of the ordered systems using the microscopic theory. According to this formalism, to find the rigidity one should introduce fictitious electric charge q for the helium-4 atoms to couple the system with electromagnetic field vector potential $\vec{A}$. In the final expression for the mass current the charge q will be reduced to zero. The advantage of introducing the electric charge is that in an electrically charged system, such as electronic liquid in metals, an additional symmetry of physical laws takes place, the local gauge symmetry, as distinct from the global gauge symmetry of a neutral liquid. According to this local symmetry the energy of the system is invariant not only under global gauge transformation, but also under a space-dependent transformation. The latter changes both the phase of the wave function and the vector potential of the electromagnetic field:

$$\Phi \to \Phi + \alpha(\vec{r}) , \quad \vec{A} \to \vec{A} + \frac{\hbar}{q}\vec{\nabla}\alpha . \tag{3.3}$$

In order to satisfy this local symmetry, Eq. (3.2) should be modified to include the compensating electromagnetic field. This is achieved by the substitution $\vec{v}_s \rightarrow \vec{v}_s - \frac{q}{m_4}\vec{A}$. So the gauge invariant London energy for the charged superfluid ^{4}He (and also for the superconductors, where one should substitute $q \rightarrow 2e$ and $m_4 \rightarrow 2m_e$, with e and m_e being the electric charge and mass of the electron respectively) is:

$$F_{\text{grad}}^{\text{London}} = \frac{1}{2}\rho_s(\vec{v}_s - \frac{q}{m_4}\vec{A})^2 . \tag{3.4}$$

The variation over $\vec{A}$ gives the electric current and therefore the mass current, which is m_4/q times electric current:

$$\vec{j} = \frac{m_4}{q}\frac{\delta F}{\delta \vec{A}} = \rho_s(\vec{v}_s - \frac{q}{m_4}\vec{A}) . \tag{3.5}$$

In the limit of zero charge q this transforms to the finite value $\vec{j} = \rho_s\vec{v}_s$ of the supercurrent.

Equation (3.5) provides the method to calculate the generalized rigidity ρ_s from the microscopic theory. One must just calculate the linear response of the mass current to the external gauge field potential:

$$(\rho_s)_{ij} = (m_4/q)\delta j_i/\delta A_j .$$

It is important that only the rotational part of the gauge potential, i.e., such $\vec{A}$ that $\vec{\nabla} \times \vec{A} \neq 0$, contributes to the supercurrent, since the curl-free field $\vec{A} = \vec{\nabla}\chi$ can be eliminated by the local gauge transformation (3.3) with $\alpha = \chi$. In the Fourier transform this means that only the transverse to the momentum $\vec{q}$ component of the vector potential, $\vec{A}_{\vec{q}} \perp \vec{q}$, contributes to the current. Therefore this is usually called the response of the system to the transverse gauge potential. The nonzero response manifests the spontaneous breaking of the $U(1)$ symmetry.

To calculate the rigidity in the systems with other broken symmetries, such as $SO_3^{(S)}$ and $SO_3^{(L)}$, one should find the response on the other local gauge fields, which are generally referred to as Yang-Mills fields. For the spin-ordered systems, where the group of the spin rotations $SO_3^{(S)}$ is broken, one should use the $SU(2)$ Yang-Mills fields, while for the crystals and liquid crystals as well as for the orbital degrees of freedom in superfluid ^{3}He, where the Euclidean group is broken, one should use the gravity fields which are related to the local Euclidean group of the general coordinate transformations. As we shall see below there are 9 rigidity parameters in the A-phase and 3 in the B-phase.

3.4. Nonpotential Superflow in ³He-A

The supercurrent in superfluid ³He is different for the B and A phases, which can be seen from Eqs. (2.9), (2.10) for their equilibrium order parameters. In the B-phase the condensate phase Φ is one of the degeneracy parameters, as a result the supercurrent has the same expression as for ⁴He, with the only one modification – in the expression for the superfluid velocity $\vec{v}_s$ the helium-4 atom mass m_4 should be substituted for by the boson mass (Cooper pair mass), i.e. $M = 2m_3$:

$$\vec{j}(\text{B-phase}) = \rho_s \vec{v}_s , \quad \vec{v}_s = \frac{\hbar}{M}\vec{\nabla}\Phi . \tag{3.6}$$

This superflow is curl-free as in superfluid ⁴He, $\vec{\nabla} \times \vec{v}_s = 0$.

In the A-phase the situation is very unusual. In addition to the anisotropy, which makes the density of the superfluid component to be a tensor (see Eqs. (2.16), (2.17)), the order parameter does not contain the phase Φ explicitly. Due to the broken relative gauge-orbital symmetry the superfluidity is manifested here, not by the phase of the complex scalar, but by the complex vector $\hat{e}^{(1)} + i\hat{e}^{(2)}$ in Eq. (2.10) or by the solid angle of the orbital rotation of the *dreibein* $\hat{e}^{(1)}$, $\hat{e}^{(2)}$ and $\hat{l}$. Taking into account that the θ_3 rotation about axis $\hat{l}$ has the same property as the gauge transformation, one may define the superfluid velocity in terms of the gradient of θ_3. As a result one obtains the equation for the superfluid velocity in terms of the degeneracy parameters, which satisfies all the symmetries of the A-phase state and properly changes under the Galilean transformation $\hat{e}^{(1)} + i\hat{e}^{(2)} \rightarrow (\hat{e}^{(1)} + i\hat{e}^{(2)}) \exp(iM\vec{w} \cdot \vec{r}/\hbar)$:

$$\vec{v}_s = -\frac{\hbar}{M}\vec{\nabla}\theta_3 = \frac{\hbar}{M}e_i^{(1)}\vec{\nabla}e_i^{(2)} . \tag{3.7}$$

What is most important here is that the vorticity of superflow, $\vec{\nabla} \times \vec{v}_s$, is not identically zero, unlike that in superfluid ⁴He and ³He-B. Instead, it depends on the $\hat{l}$ texture by virtue of the celebrated Mermin-Ho relation, which is obtained by applying the *curl* operation to both sides of Eq. (3.7):

$$\vec{\nabla} \times \vec{v}_s = -\frac{\hbar}{M}\vec{\nabla} \times \vec{\nabla}\theta_3 = \frac{\hbar}{M}\vec{\nabla}\theta_1 \times \vec{\nabla}\theta_2 , \qquad (3.8a)$$

where θ_1 and θ_2 are rotations about $\hat{e}^{(1)}$ and $\hat{e}^{(2)}$ respectively and we have used the noncommutativity of the solid rotations. The right-hand side of Eq. (3.8a) may be rewritten in terms of $\hat{l}$, since the variation of $\hat{l}$ is expressed in terms of the angle $\vec{\theta}$ variation: $\delta\hat{l} = \delta\vec{\theta} \times \hat{l}$ and $\nabla_i\hat{l} = \nabla_i\vec{\theta} \times \hat{l}$. As a result one comes to the Mermin-Ho relation:

$$\vec{\nabla} \times \vec{v}_s = \frac{\hbar}{2M}\,e_{ikl}\,l_i\vec{\nabla}l_k \times \vec{\nabla}l_l , \qquad (3.8b)$$

which intrinsically couples the superfluid properties of the A-phase with the texture in the liquid-crystal field $\hat{l}$.

So as a rule the A-phase superflow is nonpotential in the presence of the $\hat{l}$ texture. The superflow is irrotational only within such texture, at which the right-hand side of Eq. (3.8b) is exactly zero: a planar $\hat{l}$ field with the $\hat{l}$ vector constrained to be in some plane is an example of $\hat{l}$ texture with irrotational superflow.

The nonpotential character of the superfluid velocity reflects the fact that the gauge group becomes coupled with the non-Abelian group of space rotations $SO_3^{(L)}$ through the combined gauge-orbital symmetry.

3.5. *Perpetuum Motion of the A-phase*

An example of the nonpotential superflow is given by the $\hat{l}$ texture in the spherical vessel. For the axially symmetric texture the Mermin-Ho relation is simplified

$$\vec{\nabla} \times \vec{v}_s = \frac{\hbar}{2m_3\rho}(\hat{z}\partial_\rho l_z - \hat{\rho}\partial_z l_z) . \qquad (3.8c)$$

Here $\hat{\phi}$, $\hat{\rho}$ and $\hat{z}$ are unit vectors of the cylindrical coordinate system. It is evident that of all the three textures in Fig. 3.1, the $\hat{z}$ projection of the $\hat{l}$ vector, l_z, cannot avoid depending on ρ or z, therefore these textures are always accompanied by superflow. Note that the texture in Fig. 3.1c corresponds to the ground state of the system in the vessel. So the A-phase in the vessel exhibits spontaneous superflow representing the perpetuum mobile. Of course one cannot make use of the kinetic energy of the flow, since this flow corresponds to the state with minimal energy.

Equation (3.8c) can be integrated to obtain the velocity distribution in terms of the $\hat{l}$ texture. The resulting velocity field chosen i) to satisfy boundary condition, $\vec{j} \cdot \hat{\nu} = 0$, which means that there is no mass current into the wall of the container, ii) to have no singularity in the bulk liquid, and iii) to be axially symmetric, is as follows:

$$\vec{v}_s = \hat{\phi}\frac{\hbar}{2m_3\rho}(1 - l_z) \ . \tag{3.8d}$$

There is no singularity on the axis (at $\rho = 0$) in spite of the vanishing denominator: since $l_z(\rho = 0) = 1$, the dangerous denominator is compensated for. So in the ground state of the A-phase in the spherical vessel the superflow is circulating around some preferred axis $\hat{z}$, which indicates the position of the point defect, boojum, on the surface of the container in Fig. 3.1c or the orientation of the vortex lines in Figs. 1a,b. The appearance of the preferred axis in the spherically symmetric vessel is another manifestation of the broken relative gauge-orbital symmetry in the A-phase. If only the $SO_3^{(L)}$ symmetry is broken with only the $\hat{l}$ vector being the degeneracy parameter, then one can construct the axially symmetric state in the spherically symmetric vessel: $\hat{l}(\vec{r}) = \hat{r}$. But in our case of broken relative gauge-orbital invariance this spherically symmetric $\hat{l}$ texture should be accompanied by the spherically non-symmetric superflow.

3.6. *Textural Energy and Supercurrent in ^3He-A*

The total gradient energy of the A-phase in the London limit includes, besides the superfluid velocity $\vec{v}_s$, also other gradients of the degeneracy parameters in Eq. (2.10). To satisfy the invariance of $F_{\text{grad}}^{\text{London}}$ under the group G one should organize the gradients of the orbital complex vector $\hat{e}^{(1)} + i\hat{e}^{(2)}$ in physical combinations, which are invariant under the global gauge transformation $U(1)$. These are $\vec{v}_s$ and the gradients of $\hat{l}$. Therefore, the general expression for the A-phase energy in the London limit, which proves to be valid throughout the whole A-phase region since it is defined completely by the symmetries H and G, is as follows:

$$F_{\text{grad}}^{\text{London}} = \frac{1}{2}(\rho_s)_{ij}(\vec{v}_s)_i(\vec{v}_s)_j + \frac{1}{2}K_{ijmn}\partial_i\hat{l}_m\partial_j\hat{l}_n$$

$$+C_{ij}(\vec{v}_s)_i(\vec{\nabla}\times\hat{l})_j + \frac{1}{2}(\rho_{\text{sp}})_{ij}\nabla_i\hat{d}_\alpha\nabla_j\hat{d}_\alpha . \qquad (3.9a)$$

The first term in Eq. (3.9a) is the kinetic energy of superflow for the anisotropic superfluids, discussed above. In addition to that term one should also write the kinetic energy of normal flow

$$\frac{1}{2}(\rho_n)_{ij}(\vec{v}_n)_i(\vec{v}_n)_j ,$$

with

$$(\rho_n)_{ij} + (\rho_s)_{ij} = \rho\delta_{ij} ,$$

where ρ is the total mass density of the liquid.

The second term in Eq. (3.9a) is the energy of the liquid-crystal field, $\hat{l}$ distortion, which contains three independent rigidity parameters

$$\frac{1}{2}K_{ijmn}\partial_i\hat{l}_m\partial_j\hat{l}_n = \frac{1}{2}\left(K_1(\vec{\nabla}\cdot\hat{l})^2 + K_2(\hat{l}\cdot(\vec{\nabla}\times\hat{l}))^2 + K_3(\hat{l}\times(\vec{\nabla}\times\hat{l}))^2\right) . \quad (3.9b)$$

Here K_1, K_2, and K_3 are the twist, splay, and bend coefficients respectively, as in nematic liquid crystals.

The third term in Eq. (3.9a)

$$C_{ij}(\vec{v}_s)_i(\vec{\nabla}\times\hat{l})_j = C\,\vec{v}_s\cdot(\vec{\nabla}\times\hat{l}) - C_0(\hat{l}\cdot\vec{v}_s)(\hat{l}\cdot(\vec{\nabla}\times\hat{l})) , \qquad (3.9c)$$

originates from the interaction of the $\vec{v}_s$ field with the orbital degrees of freedom.

The supercurrent $\vec{j}$ in ^{3}He-A is obtained again by introducing the compensating electromagnetic field, and contains two different terms, $\vec{j} = \vec{j}_s + \vec{j}_{\text{orb}}$, where $\vec{j}_s$ describes an anisotropic superflow with the superfluid velocity $\vec{v}_s$:

$$\vec{j}_s = \rho_s^{\|}\hat{l}(\hat{l}\cdot\vec{v}_s) + \rho_s^{\perp}\left(\vec{v}_s - \hat{l}(\hat{l}\cdot\vec{v}_s)\right) . \qquad (3.10)$$

Another current which comes from Eq. (3.9c) is orbital:

$$\vec{j}_{\text{orb}} = C\vec{\nabla}\times\hat{l} - C_0\hat{l}\left(\hat{l}\cdot(\vec{\nabla}\times\hat{l})\right) . \qquad (3.11)$$

It is produced by internal rotations of Cooper pairs around $\hat{l}$, the currents produced by these rotations do not compensate each other if the $\hat{l}$ texture is inhomogeneous.

3.7. Spin Soliton in ^{3}He-A

Finally the fourth term in Eq. (3.9a) describes the energy of the inhomogeneity of the spin part $\hat{d}$ of the order parameter. The spin rigidity parameter is a uniaxial tensor with the anisotropy axis along the orbital vector $\hat{l}$:

$$(\rho_{sp})_{ij} = \rho_{sp}^{\parallel} \hat{l}_i \hat{l}_j + \rho_{sp}^{\perp}(\delta_{ij} - \hat{l}_i \hat{l}_j) \,, \qquad (3.12)$$

with $\rho_{sp}^{\parallel}$ and $\rho_{sp}^{\perp}$ referring to the longitudinal and transverse components of the spin rigidity. This term, together with the spin-orbital coupling energy (2.18) corresponds to the energy of two-sublattice antiferromagnets with the antiferromagnetic axis $\hat{d}$ which has an easy-axis "crystal lattice" anisotropy along the $\hat{l}$ vector.

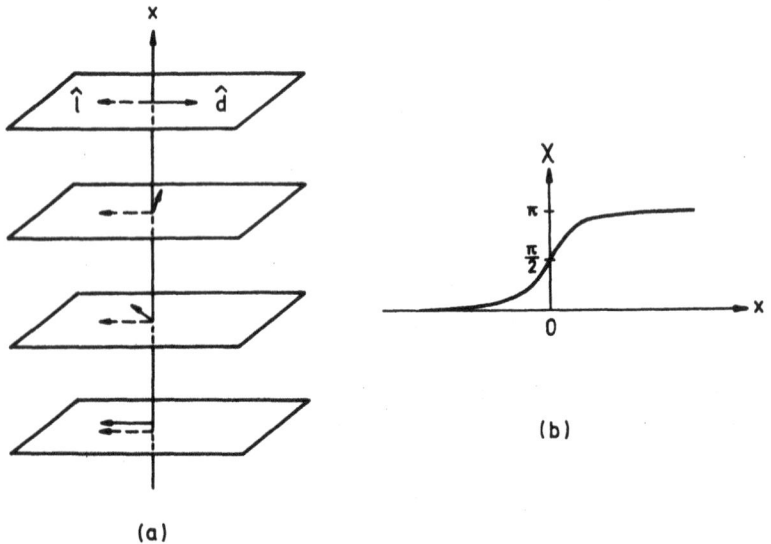

Fig. 3.3. $\hat{d}$ soliton in the A-phase. a) $\hat{d}$ performs twisting motion from the parallel to antiparallel orientation with respect to $\hat{l}$. b) The twisting angle $\chi(x)$.

As an example of the texture which exists in the London limit, let us consider the $\hat{d}$ soliton, the metastable domain wall in which the degeneracy parameter $\hat{d}$ changes from one equilibrium position $\hat{d} = \hat{l}$ on one side of the wall to another equilibrium position $\hat{d} = -\hat{l}$ on the other side.

If one neglects the change in the orbital degeneracy parameter $\hat{e}^{(1)} + i\hat{e}^{(2)}$, this texture is defined by the following free energy functional:

$$F_{\text{soliton}} = \frac{1}{2} \int dx \left(\rho_{\text{sp}}^{\perp} (\nabla_x \hat{d})^2 - g_{\text{so}} (\hat{d} \cdot \hat{z})^2 \right) , \qquad (3.13)$$

where we have chosen the axis x along the normal to the soliton wall, and the direction of $\hat{l}$ along z. The second term in Eq. (3.13) is a small spin-orbit coupling in Eq. (2.18) which fixes the orientation of $\hat{d}$ parallel or antiparallel to $\hat{l}$ outside the soliton. The solution of the corresponding equation

$$\xi_{d\perp}^2 \partial_x^2 \chi = \frac{1}{2} \sin 2\chi , \qquad (3.14)$$

for the angle χ between $\hat{d}$ and $\hat{l}$ vectors ($\hat{d} = \hat{z} \cos \chi(x) + \hat{y} \sin \chi(x)$) gives the following $\hat{d}$ texture in the soliton:

$$\tan \frac{\chi}{2} = \exp \frac{x}{\xi_{d\perp}} . \qquad (3.15)$$

In this solution the angle χ changes from 0 at $x = -\infty$ to π at $x = +\infty$, and these equilibrium asymptotic values are achieved at $|x| \gg \xi_{d\perp}$, where $\xi_{d\perp} = \sqrt{\rho_{\text{sp}}^{\perp}/g_{\text{so}}}$ is the so-called dipole length, of order 10^{-3}cm, which defines the thickness of the soliton wall. This thickness is much larger than the characteristic coherence length $\xi \sim 10^{-5}$–10^{-6}cm, so the condition for the London limit is satisfied. Such kind of solitons is observed in NMR experiments on superfluid ^{3}He-A.

3.8. *Order Parameter Textures*

Above we have considered textures in the field of the degeneracy parameters. Since the degeneracy parameters do not depend on temperature, the London gradient energy of inhomogeneity is valid for all temperatures below T_c, with the rigidity parameters being the functions of the temperature and pressure. This London energy defines the structure of the textures, provided that the condition for the London limit is satisfied. However some textures exist where this condition is violated, for example in the core of quantized vortex in superfluid ^{4}He and ^{3}He-B the order parameter no longer corresponds to the equilibrium value and therefore is no longer described by the degeneracy parameters only. In this case one should extend the description to include the other, non-Goldstone, degrees of freedom.

This may be easily done near T_c where the Ginzburg-Landau theory can be applied, which operates with the 18 components of the order parameter $A_{\alpha i}$ instead of the 5 Goldstone variables (degeneracy parameters) in the A-phase and 4 in the B-phase. To describe the order parameter textures one should add to the bulk energy in Eq. (2.1) an energy in terms of the order parameter gradients, which has the following general form dictated by the G symmetry:

$$F_{\text{grad}}^{\text{G-L}} = \gamma_1 \partial_i A_{\alpha j} \partial_i A_{\alpha j}^* + \gamma_2 \partial_i A_{\alpha i} \partial_j A_{\alpha j}^* + \gamma_3 \partial_i A_{\alpha j} \partial_j A_{\alpha i}^* . \qquad (3.16)$$

With the energy functional $F_{\text{bulk}}^{\text{G-L}} + F_{\text{grad}}^{\text{G-L}}$, Eq. (3.16) + Eq. (2.1), one may for example, study the structure of the core of the B-phase vortex and identify a textural phase transition within the core, which has been experimentally observed in the Helsinki NMR experiments on rotating ^{3}He-B, as a breaking of axisymmetry of the vortex core.

3.9. *Coherence Length and London Limit*

Now we can discuss the question in which cases one should use the Ginzburg-Landau free energy functional, Eq. (2.1) + Eq. (3.16), for the whole order parameter $A_{\alpha i}$, or the London free energy functional for the degeneracy parameters of a given superfluid phase (Eq. (3.9a) in the case of the A-phase). The superfluid phases are defined as minima of the bulk term $F_{\text{bulk}}^{\text{G-L}}$ of the Ginzburg-Landau functional, so the inhomogeneity does not perturb the superfluid state essentially if $F_{\text{grad}}^{\text{G-L}} \ll F_{\text{bulk}}^{\text{G-L}}$. If one compares these two free energy terms, one can see that there exists a combination of the parameters α and γ's, $\xi \sim \sqrt{\alpha/\gamma}$, which has the dimension of length and is called the coherence length (more exactly there are two coherence lengths, which are of the same order of magnitude: $\xi_\perp = \sqrt{\alpha/\gamma_1}$, and $\xi_\parallel = \sqrt{\alpha/(\gamma_1 + \gamma_2 + \gamma_3)}$). Now if the characteristic scale, l_{texture}, of the texture is large, $l_{\text{texture}} \gg \xi$, then $F^{\text{G-L}}_{\text{grad}} \ll F^{\text{G-L}}_{\text{bulk}}$, and locally in the texture one has a well defined superfluid phase and may use the London free energy functional for the degeneracy parameters of this phase.

To obtain the London energy near T_c, for example, for the A-phase one should insert the A-phase order parameter (Eq. (2.10)) into $F^{\text{G-L}}_{\text{grad}}$:

$$F^{\text{London}}_{\text{grad}}(\text{A-phase}) = F^{\text{G-L}}_{\text{grad}}\{A_{\alpha i} = A_{\alpha i}(\text{A-phase})\} . \tag{3.17}$$

Here we consider the simplest case of the Bardeen-Cooper-Schrieffer (BCS) model of superfluid ^{3}He, also known as weak coupling approximation. Within this model all three parameters in the gradient energy are equal, $\gamma_1 = \gamma_2 = \gamma_3 = \gamma$). As a result one obtains the following values for 9 rigidity parameters in $F^{\text{London}}_{\text{grad}}$ for the A-phase in the temperature region close to T_c:

$$\left(\frac{\hbar}{M}\right)^2 \rho^{\|}_s = \rho^{\|}_{\text{sp}} = \frac{1}{2}\left(\frac{\hbar}{M}\right)^2 \rho^{\perp}_s = \frac{1}{2}\rho^{\perp}_{\text{sp}} = 2K_1 = 2K_2$$

$$= \frac{2}{3}K_3 = \left(\frac{\hbar}{M}\right)C_0 = 2\left(\frac{\hbar}{M}\right)C$$

$$= 4\gamma\Delta^2_A(T) . \tag{3.18}$$

3.10. *Disgyrations and Vortex. London Equations for the Orbital Texture*

To show examples of texture where one should apply the total Ginzburg-Landau free energy functional, let us consider three linear objects: quantized vortex and the so-called radial (Fig. 3.4a) and tangential (Fig. 3.4b) disgyrations in the A-phase.

The disgyration is the linear object with a nonzero winding number m_l for the $\hat{l}$ vector around the line: $\hat{l} = \pm\hat{\rho}$ for the radial disgyration and $\hat{l} = \pm\hat{\phi}$ for the tangential disgyration. Here $\hat{z}$, $\hat{\rho}$, and $\hat{\phi}$ are unit vectors of cylindrical coordinate frame with $\hat{z}$ along the line of defect. For all the four disgyrations discussed here the winding number $m_l = 1$: i.e., the $\hat{l}$ vector performs 2π rotation about the disgyration axis in the positive direction.

The vortex is the linear object with a nonzero winding number for the angle θ_3 of the rotation of the *dreibein* about axis $\hat{l}$. We further denote

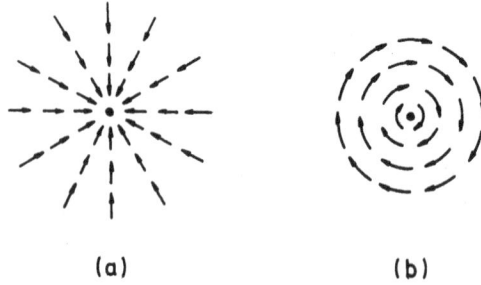

(a)   (b)

Fig. 3.4. Linear defects in the A-phase: radial ($\hat{l} = -\hat{\rho}$ in (a)) and tangential ($\hat{l} = -\hat{\phi}$ in (b)) disgyrations. The vector $\hat{l}$ is not well defined on the axis of the defect, where the superfluid is no more in the A-phase state.

the θ_3 winding number as m_Φ, i.e. the phase Φ winding number, since θ_3 rotation of the *dreibein* corresponds to a gauge transformation. In the simplest case of the pure vortex one has $\hat{l} = \hat{z}$ and $\theta_3 = m_\Phi \phi$.

All these textures represent exact solutions of the Euler-Lagrange equation obtained from the London energy functional, Eq. (3.9a), for the $\hat{l}$ and $\vec{v}_s$ (or θ_3) fields:

$$\hat{l} \times \frac{\delta F_{\text{grad}}^{\text{London}}}{\delta \hat{l}} = 0 \,, \quad \frac{\delta F_{\text{grad}}^{\text{London}}}{\delta \theta_3} = 0 \,. \tag{3.19}$$

Here we have neglected the spin-orbit interaction, i.e., we consider the length scale smaller than the dipole length ξ_d, so the gradient energy is larger than the spin-orbital one. In the first of Eqs. (3.19) one should take into account that $\vec{v}_s$ depends on $\hat{l}$ through the Mermin-Ho relation, thus

$$\frac{\delta F_{\text{grad}}^{\text{London}}}{\delta \hat{l}} = \left(\frac{\delta F_{\text{grad}}^{\text{London}}}{\delta \hat{l}}\right)_{v_s} + \left(\frac{\delta F_{\text{grad}}^{\text{London}}}{\delta \vec{v}_s}\right)_l \frac{\delta \vec{v}_s}{\delta \hat{l}} = 0 \,, \tag{3.20a}$$

$$\frac{\delta(\vec{v}_s)_i}{\delta \hat{l}} = \frac{\hbar}{2m_3} \hat{l} \times \nabla_i \hat{l} \,. \tag{3.20b}$$

The second of Eqs. (3.19) is the continuity equation for the mass current:

$$\vec{\nabla} \cdot \vec{j} = 0 \,, \tag{3.20c}$$

since

$$\frac{\delta F_{\text{grad}}^{\text{London}}}{\delta \theta_3} = -\vec{\nabla} \cdot \frac{\partial F_{\text{grad}}^{\text{London}}}{\vec{\nabla} \theta_3} = \frac{2m_3}{\hbar} \vec{\nabla} \cdot \frac{\partial F_{\text{grad}}^{\text{London}}}{\delta \vec{v}_s} = \frac{2m_3}{\hbar} \vec{\nabla} \cdot \vec{j} \; .$$

In the simplest case of the disgyrations and vortices discussed here, the $\hat{l}$ field is constrained in the plane, so the right-hand side of the Mermin-Ho relation is exactly zero, as a result $\vec{v}_s$ (or θ_3) field is decoupled from the $\hat{l}$ field. So one has the separate equation (3.20a) for the $\hat{l}$ field in disgyrations with $\vec{v}_s = 0$, and separate equation (3.20c) for the θ_3 texture in the vortex with the constant field $\hat{l} = \hat{z}$.

The equation (3.20a) for the planar $\hat{l}$ texture is thus a closed equation with only the terms in Eq. (3.9b) for $F_{\text{grad}}^{\text{London}}$ taken into account:

$$K_3 \nabla^2 \hat{l} + (K_1 - K_3)(\hat{l} \cdot \vec{\nabla})(\vec{\nabla} \cdot \hat{l}) = 0 \; . \tag{3.21a}$$

Here both $\vec{\nabla}$ and $\hat{l}$ are assumed to be constrained in the (x, y) plane, since we are looking for the planar texture, $\hat{l} \perp \hat{z}$, which in the case of linear object depends only on two coordinates, x and y. This equation has two axisymmetric solutions: the radial disgyration $\hat{l} = \pm\hat{\rho}$ and the tangential disgyration $\hat{l} = \pm\hat{\phi}$.

Equation (3.20c) which describes the vortex is also a closed equation

$$\nabla^2 \theta_3 = 0 \; . \tag{3.21b}$$

It has the following axisymmetric solutions $\theta_3 = m_\Phi \phi$, corresponding to the quantum vortices with the winding number m_Φ. The superfluid velocity in these vortices

$$\vec{v}_s = m_\Phi \frac{\hbar}{2m_3} \frac{\hat{\phi}}{\rho} \; , \tag{3.22a}$$

is circulating about the vortex axis with the circulation

$$\oint d\vec{r} \cdot \vec{v}_s = m_\Phi \frac{h}{2m_3} \; , \tag{3.22b}$$

being quantized in terms of the elementary circulation quantum $\kappa = \frac{h}{2m_3}$.

3.11. *Disgyrations and Vortex. Singularity in the Degeneracy Parameters*

For all three linear defects the degeneracy parameter, $\hat{l}$ for disgyrations and θ_3 for vortices, is ill defined at the origin ($\rho = 0$), i.e., the degeneracy parameter field has a singularity at $\rho = 0$. Also the London energy diverges at the origin: for the radial disgyration the elastic energy of the $\hat{l}$ field, $F_{\text{grad}}^{\text{London}} = K_1/2\rho^2$, is infinite at $\rho = 0$; for the quantized vortex the kinetic energy of superflow $F_{\text{grad}}^{\text{London}} = (\rho_s)_\perp m_\Phi^2(\kappa/2\pi)^2/2\rho^2$ is infinite at $\rho = 0$. The total energy of the linear object (per unit length) is logarithmically divergent:

$$F_{\text{vortex}} = \int d^2\rho \; F_{\text{grad}}^{\text{London}} = \pi(\rho_s)_\perp m_\Phi^2 (\frac{\kappa}{2\pi})^2 \ln \frac{R}{\xi} , \qquad (3.23a)$$

$$F_{\text{radial disgyration}} = \int d^2\rho \; F_{\text{grad}}^{\text{London}} = \pi K_1 \ln \frac{R}{\xi} , \qquad (3.23b)$$

$$F_{\text{tangential disgyration}} = \int d^2\rho \; F_{\text{grad}}^{\text{London}} = \pi K_3 \ln \frac{R}{\xi} , \qquad (3.23c)$$

where R is the external cutoff parameter, which is either the size of the vessel or the distance to a neighbouring singularity.

The lower cutoff parameter is the coherence length $\xi \sim \sqrt{K/\alpha\Delta_A^2}$, the distance at which the London approximation becomes invalid, since the gradient energy becomes comparable with the bulk one. Below this distance, in the hard core of defect, one should look for a more general solution of the total G-L equations, including the gradient terms, for the order parameter $A_{\alpha i}$, which is no more on the A-phase manifold. As distinct from $\hat{l}$ and θ_3, the order parameter $A_{\alpha i}$ should be defined everywhere.

This is one of the most important differences between the degeneracy parameter and the order parameter. This comes from the fact that the first varies on the compact manifold (circumference, sphere, the space of the *dreibein*, or more complicated manifolds in Eqs. (2.20), (2.21)), for example in our case $\hat{l}$ is constrained to be a unit vector. This constraint is sometimes incompatible with the given asymptotic distribution of the degeneracy parameter field or with the boundary conditions. As a result the degeneracy parameter is not defined at some points, lines or surfaces, which are called singular points, lines or surfaces. Such singularities are investigated by topological methods.

On the other hand, the order parameter $A_{\alpha i}$ is defined on a simple R^{18} space without any constraints, therefore it is well defined everywhere in space, even in places where the degeneracy parameters have singularities, i.e., in the hard cores of disgyrations and vortices.

3.12. *Radial Disgyration. The Hard Core Structure*

To find the order parameter field within the hard core one should solve the general G-L equations $\frac{\delta F}{\delta A_{\alpha i}} = 0$. The simplest, i.e. the most symmetric, solution for the radial disgyration is as follows:

$$A_{\alpha i}(\rho, \phi) = \Delta_A \hat{z}_\alpha (a(\rho)\hat{z}_i + ib(\rho)\hat{\phi}_i) , \qquad (3.24a)$$

where $a(\rho)$ and $b(\rho)$ are real functions satisfying the equations

$$\xi_\perp^2 (\partial_\rho^2 a + \frac{1}{\rho}\partial_\rho a) + 2a - a(a^2 + b^2) - \frac{\beta_{13}}{\beta_{245}}a(a^2 - b^2) = 0 , \qquad (3.24b)$$

$$\xi_\perp^2 (\partial_\rho^2 b + \frac{1}{\rho}\partial_\rho b - \frac{1}{\rho^2}b) + 2b - b(a^2 + b^2) - \frac{\beta_{13}}{\beta_{245}}b(b^2 - a^2) = 0 , \qquad (3.24c)$$

which can be solved numerically.

Both $a(\rho)$ and $b(\rho)$ tend to unity at $\rho \gg \xi$, where the order parameter corresponds to the A-phase state, Eq. (2.10), with $\hat{d} = \hat{e}^{(1)} = \hat{z}$, $\hat{e}^{(2)} = \hat{\phi}$, and with the disgyration in the $\hat{l}$ field: $\hat{l} = \hat{\rho}$. However in the region $\rho \leq \xi$ the functions $a(\rho)$ and $b(\rho)$ are no longer unity, and moreover are no longer equal. Since the vector ϕ is not defined at the origin the function $b(\rho)$ should vanish on the axis of disgyration: $b(0) = 0$; for the function $a(\rho)$ such restriction is absent, $a(0) \neq 0$. So at the origin the order parameter,

$$A_{\alpha i}(\rho = 0) = \Delta_A a(0)\hat{z}_\alpha \hat{z}_i , \qquad (3.25)$$

is well defined and is nonzero. It corresponds to another possible superfluid phase, which however does not exist in the bulk liquid due to unfavorable β parameters in the G-L functional. This is the so-called polar phase. The region within the radius $\rho \leq \xi$, where there is no pure A-phase state and due

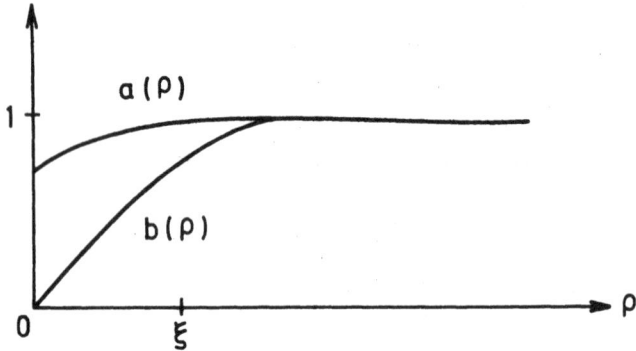

Fig. 3.5. Structure of the hard core of the radial disgyration in the A-phase. Singularity on the axis is smoothed out not by violation of the superfluidity on the axis but by introducing a new superfluid phase in the vortex core. In the region of the hard core, $\rho \sim \xi$, the components a and b of the order parameter are split and the A-phase transforms to the polar phase (Eq. (3.25)) on the axis of disgyration.

to this the singularity in the degeneracy parameter is resolved, is called the core of the disgyration, or more generally the hard core of the topological defect.

3.13. *Pure Vortices. The Hard Core Structure*

In the same manner the solution of the London equation (3.21b) for the vortex

$$A_{\alpha i}^{\text{London}}(\rho, \phi) = \Delta_A \hat{z}_\alpha(\hat{x}_i + i\hat{y}_i)e^{i\phi m_\Phi} , \qquad (3.26a)$$

is badly defined at the origin, where the azimuthal angle ϕ is not defined. So in the region of the hard core this solution of the London equation should be modified to become the solution of the G-L equations. The simplest, i.e. the most symmetric, solution for the order parameter in the hard core of the quantized vortex with m_Φ circulation quanta and with $\hat{l} = \hat{z}$ is also expressed in terms of two independent parameters, the amplitudes C_{0+} and C_{0-} of the Cooper pair states with given projections of spin and orbital momenta:

$$A_{\alpha i}(\rho, \phi) = \Delta_A \hat{z}_\alpha \left(C_{0+}(\rho)(\hat{x}_i + i\hat{y}_i)e^{i\phi m_\Phi} + C_{0-}(\rho)(\hat{x}_i - i\hat{y}_i)e^{i\phi(m_\Phi+2)} \right) .$$
$$(3.26b)$$

The behavior of these parameters is shown in Fig. 3.6 for the case $m_\Phi = -2$. Outside the hard core only the amplitude C_{0+} is present, which corresponds to the initial A-phase state with $\hat{l} = \hat{z}$ and with $\theta_3 = m_\Phi \phi$ in Eq. (3.26a). In the hard core a new amplitude C_{0-} appears, which corresponds to the A-phase state with an opposite orbital momentum: $\hat{l} = -\hat{z}$. This new A-phase state has a different winding number; $m_\Phi \to m_\Phi + 2$, while $M_L + m_\Phi$ is equal for both components (this relation means the axial symmetry of the vortex, as will be discussed in Sec. 8.3). Both amplitudes C_{0+} and C_{0-} vanish at the origin, where the azimuthal angle ϕ is not defined, since each term in Eq. (3.26b) has nonzero winding numbers: m_Φ and $m_\Phi + 2$ correspondingly.

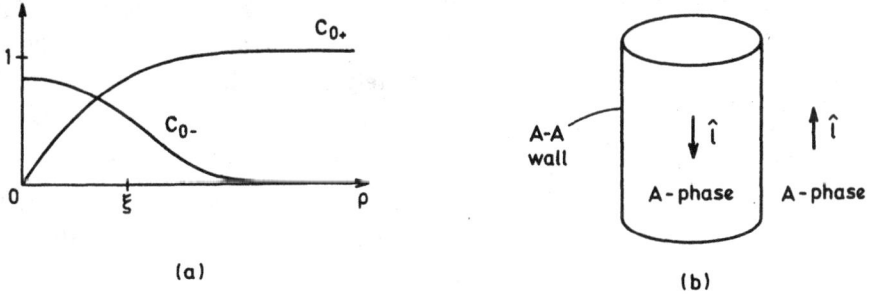

Fig. 3.6. Core structure of the most symmetric vortex with winding number $m_\Phi = -2$ in the A-phase. Both outside the hard core and on the vortex axis the liquid is in the superfluid A-phase states with two different orientations of the $\hat{l}$ vector. The intermediate region represents the domain wall between two A-phase states (b), within the A-A wall the system is no longer in the A-phase manifold.

The only exception from this rule is that for the vortex with $m_\Phi = -2$, for which the winding number is zero in the second term in Eq. (3.26b), therefore the amplitude C_{0-} in Eq. (3.26b) is finite on the axis of this vortex. The hard core of the vortex with $m_\Phi = -2$ is thus superfluid and consists of the homogeneous A-phase state with the opposite direction of $\hat{l}$ but without vortex singularity. Two A-phase states are separated by the domain wall

(Fig. 3.6b) which represents a vortex sheet. The vortex singularity on the axis is thus transformed into the vortex sheet within which the superfluid state is no more the A-phase.

Two questions arise when one considers the disgyration or quantized vortex. 1) Why is the disgyration or vortex stable? Perhaps it is possible to unwind continuously the *dreibein* field and obtain the texture without a hard core of the coherence length size. 2) Why does the solution for the 18 component order parameter contain only 2 nonzero functions for the disgyration and the vortex?

The topological classification gives an answer to the first question, which shows how to distribute defects into topologically stable classes, in particular it will be shown that disgyration and quantized vortices with odd m_Φ belong to the same nontrivial topological class, which means that these structures can be transformed into each other by continuous deformation of the *dreibein* field. However, complete unwinding is topologically impossible, so these defects are topologically stable. While the singularity in the vortices with even m_Φ can be completely unwound.

The symmetry classification of textures within a given topological class shows how many independent parameters exist for the defect of certain symmetry. This symmetry analysis of inhomogeneous states, which allows us to simplify the problem since it leads to an essential reduction of the number of components of the order parameter in the Ginzburg-Landau equations, is discussed below using the example of the AB-interface and in Sec. 8.3 for the symmetry classification of vortices.

3.14. *A-B Interface. Symmetry and Structure*

Another example of textures with hard core is the phase boundary between the ^3He-A and ^3He-B liquids. These two liquids have different symmetries, neither of the two symmetry groups is the subgroup of the other. Therefore the phase transition between them may only be of the first order. On the transition line in the pressure-temperature phase plane (see Fig. 1.1b) these two phases may coexist like water and vapour being separated by the interface. Inside this domain wall the superfluid is no more in A or in B state and therefore the general Ginzburg-Landau equations should be solved to find the exact distribution of the order parameter.

The A-B boundary which connects these two phases produces mutual boundary conditions which relate the degeneracy parameters $\hat{d}$, $\hat{e}^{(1)}$, $\hat{e}^{(2)}$ and $\vec{l}$ on the A-phase side with the degeneracy parameters Φ and $\mathbf{R}_{\alpha i}$ on the B-phase side. These boundary conditions depend on the symmetry of the order parameter within the boundary. So, again, in the same manner as the symmetry group of the superfluid state defines most of the properties of the given phase, the symmetry group $H_{\text{AB wall}}$ of the order parameter in the A-B wall defines the physical properties of the wall.

The group $H_{\text{AB wall}}$ is the subgroup of the general group G_{wall} of the physical laws in the presence of the plane, which fixes the position of some wall in the system. This wall may either separate two nonequivalent phases or be the boundary of the container, in either case G_{wall} is

$$G_{\text{wall}} = (U(1) \times T) \times C_{\infty v}^{(L)} \times SO_3^{(S)} \ . \tag{3.27}$$

The presence of the wall does not violate the gauge group $U(1)$, the time inversion symmetry T, and the spin rotation group $SO_3^{(S)}$. However the parity transformation symmetry P is violated, since the half-space on the left side of the wall is not equivalent to the right half-space. Also the orbital rotations group $SO_3^{(L)}$ is no longer the symmetry of the system in the presence of the fixed wall: only those orbital rotations are allowed which do not move the wall. So the last two groups, $P \times SO_3^{(L)}$, are reduced to the group $C_{\infty v}^{(L)}$, which contains the group $SO_2^{(L)}$ of the orbital rotations about the normal $\hat{x}$ to the wall and reflections in the planes, which are perpendicular to the plane of the wall.

The symmetry $H_{\text{AB wall}}$ of the wall structure, which includes the A-phase and B-phase states as asymptotes on both sides of the wall, should also be the subgroup of the symmetry groups of the A- and B- phases, H_A and H_B. Therefore

$$H_{\text{AB wall}} \subset (G \cap H_A \cap H_B) \ . \tag{3.28}$$

These restrictions leave not so much space for the $H_{\text{AB wall}}$. If, without loss of generality, one fixes the asymptote for the B-phase order parameter on the right-hand side of the wall as

$$A_{\alpha i}(x = +\infty) = \Delta_B \delta_{\alpha i} \ , \tag{3.29}$$

then the largest possible $H_{\text{AB wall}}$ groups contain only four elements, and there are only two nonequivalent groups of maximal symmetry:

$$H^1_{\text{AB wall}} = (1 \ , \ C^S_{\pi,x}C^L_{\pi,x}T \ , \ C^S_{\pi,y}C^L_{\pi,y}P \ , \ C^S_{\pi,z}C^L_{\pi,z}TP) \ , \qquad (3.30a)$$

$$H^2_{\text{AB wall}} = (1 \ , \ C^S_{\pi,x}C^L_{\pi,x} \ , \ C^S_{\pi,y}C^L_{\pi,y}TP \ , \ C^S_{\pi,z}C^L_{\pi,z}TP) \ . \qquad (3.30b)$$

Here $C^S_{\pi,x(y,z)}$ and $C^L_{\pi,x(y,z)}$ are π rotations about the axis $x(y,z)$ in spin and orbital spaces correspondingly.

The symmetry in Eqs. (3.30) defines the number of independent components of the order parameter. For both solutions the symmetry $H_{\text{AB wall}}$ requires that there should be exactly 5 nonzero components, for the solution with the symmetry $H^1_{\text{AB wall}}$ these are:

$$\text{Re}(A_{xx}) \ , \ \text{Re}(A_{yy}) \ , \ \text{Re}(A_{zz}) \ , \ \text{Im}(A_{zx}) \ , \ \text{Im}(A_{xz}) \ , \qquad (3.31a)$$

and for the solution with the symmetry $H^2_{\text{AB wall}}$ the nonzero components are

$$\text{Re}(A_{xx}) \ , \ \text{Re}(A_{yy}) \ , \ \text{Re}(A_{zz}) \ , \ \text{Im}(A_{zy}) \ , \ \text{Im}(A_{yz}) \ . \qquad (3.31b)$$

The numerical solution of the Ginzburg-Landau equations for the structure 1 is shown in Fig. 3.7. This structure proves to have lower energy than the structure 2 in the region of the applicability of the G-L equations for all pressures. The latter structure 2 is the metastable state of the AB interface. However it is not excluded that at lower temperatures beyond the G-L region the structure with symmetry 2 can become stable.

3.15. *Symmetry of the A-B Interface and Boundary Conditions*

The symmetry also defines the asymptotic behavior of the order parameter on the A-phase side at given order parameter asymptotics of the B-phase; for the state with symmetry 1 this is:

$$A^1_{\alpha i}(x = -\infty) = \Delta_A \hat{x}_\alpha (\hat{x}_i - i\hat{z}_i) \ , \qquad (3.32a)$$

which means that $\hat{d}$ is normal to the wall while $\hat{l}$ is parallel to the wall. For the state with symmetry 2 the A-phase asymptotic order parameter is:

$$A^2_{\alpha i}(x = -\infty) = \Delta_A \hat{y}_\alpha (\hat{y}_i + i\hat{z}_i) \ , \qquad (3.32b)$$

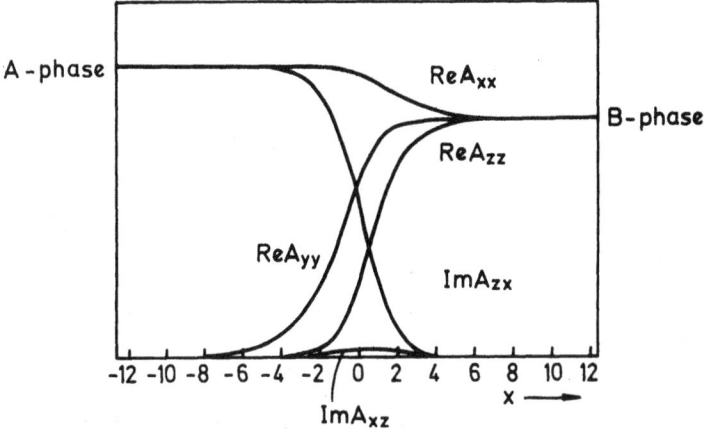

Fig. 3.7. The hard core structure of the phase boundary between A and B phases. The length is scaled in terms of the coherence length ξ.

which means that $\hat{l}$ is now normal to the wall while $\hat{d}$ is parallel to the wall.

Since we fixed the order parameter in the B-phase, Eq. (3.29), we obtained only a particular state of the AB interface. As in the case of homogeneous order parameter (see Sec. 2.5) all the degenerate states of the AB wall are obtained from the simplest ones in Eqs. (3.31) by the action of the symmetry group G_{wall}, therefore the general structure of the AB wall has the symmetry $G_{\mathrm{wall}} H_{\mathrm{AB\ wall}} G_{\mathrm{wall}}^{-1}$ and has the form (cf. Eq. (2.8)):

$$(g_{\mathrm{wall}} A^0)_{\alpha i} = \exp(i\Phi) R_{\alpha\beta}^S R_{ik}^L A_{\beta k}^0 \,, \qquad (3.33)$$

where $\mathbf{R}^S$ is the matrix of three-dimensional spin rotations as in Eq. (2.8), while $\mathbf{R}^L$ contains only orbital rotations about x.

Equation (3.33) for the group motion and Eqs. (3.29) and (3.32) for the asymptotes of the initial simplest states now define the equilibrium orientations of the degeneracy parameters on the A- and B-phase sides of the wall with respect to the wall and to each other. The asymptotes on the A-phase side and on the B-phase side are not arbitrary but are regulated by these equations, which thus give the boundary conditions on the orientation of the degeneracy parameters.

The boundary conditions may be divided into two groups: 1) The regular boundary conditions. These are the boundary conditions for the $\hat{l}$ vector of the A-phase which is oriented exclusively by the wall: $\hat{l}$ is normal to the wall for the AB interface with symmetry 2, $\hat{l}$ is parallel to the wall for the AB interface with symmetry 1, irrespective of the B-phase order parameter, and 2) mutual boundary conditions, which couple the orientations of the order parameters across the wall. An example of such boundary conditions for the wall with symmetry 1 gives the following matching rule: if the B-phase asymptote is described by the degeneracy parameter matrix $R_{\alpha i}$, then the $\hat{d}$ vector on the A-phase side is rotated by the same matrix from its initial position along the normal: $\hat{d}_\alpha = R_{\alpha i}\hat{x}_i$.

The most interesting point is that there is no one-to-one correspondence between the orientations of the order parameter across the wall; there is still some room for orientation of the degeneracy parameter on one side of the wall at fixed orientation of the degeneracy parameter on another side. For example, in the AB wall of symmetry 1 the orientation of $\hat{l}$ in the plane of the wall does not depend on the orientation of the $R_{\alpha i}$ on the B-phase side. And, vice versa, the spin rotation around axis $\hat{d}$ leaves the A-phase side invariant, but changes the B-phase order parameter. This freedom leads to the variety of the topologically stable defects, boojums and monopoles, on the AB interface.

4

Bose Excitations in Superfluid Phases of ³He

Breaking of the continuous symmetry is accompanied by appearance of 1) the order parameter in general, and 2) the degeneracy parameters (Goldstone fields) in particular. The latter describe the orientation of the order parameter in equilibrium states of a given superfluid phase. Accordingly the collective modes appear, which correspond to low frequency dynamics of different components of the order parameter. Some of these modes, which are oscillations of the Goldstone fields, are gapless (soft) Goldstone modes. The number of the Goldstone modes usually coincides with the dimension of the manifold of the degeneracy states, i.e. 5 for the A-phase and 4 for the B-phase.

4.1. Goldstone Bosons in ³He-A

In the A-phase the Goldstone bosons are : i) spin waves, which are oscillations of the unit vector of spin anisotropy: $\hat{d} = \hat{d}_0 + \delta\hat{d}$ with $\delta\hat{d} \sim \exp(i\vec{q}\cdot\vec{r} - i\omega t)$. Two polarizations of spin waves correspond to oscillations of two projections of $\delta\hat{d} \perp \hat{d}_0$. The other 3 Goldstone modes are oscillations of the orbital complex vector $\hat{e}^{(1)} + i\hat{e}^{(2)}$, these oscillations have three degrees of freedom: the three angles of solid rotations of the *dreibein* $\hat{e}^{(1)}$, $\hat{e}^{(2)}$ and $\hat{l}$. These rotations can be divided into ii) the rotation θ_3 of the vectors $\hat{e}^{(1)}$ and

$\hat{e}^{(2)}$ about vector $\hat{l}$, the corresponding Goldstone mode is analogous to the sound in ^{4}He in which the superfluid velocity $\vec{v}_s = (\hbar/2m_3)\vec{\nabla}\theta_3$ oscillates (these oscillations are manifested as 1st, 2nd, or 4th sound depending on external conditions such as geometry and temperature); and iii) rotation of the axis $\hat{l}$ itself, the corresponding oscillations of the orbital $\hat{l}$ vector with two polarizations are known as orbital waves.

The low frequency dynamics of the system with broken symmetry includes all hydrodynamical soft variables: i) Goldstone variables $\hat{d}$, $\hat{l}$ and $\vec{v}_s$; and ii) those which exist in normal liquid and which are the soft variables due to the conservation law for the corresponding quantities: mass density ρ, momentum density $\vec{p}$, entropy density S and the spin density $\vec{S}$. Since the orbital dynamics contains a number of anomalies related with the chiral anomaly in quantum field theory, we shall consider it later in a special section. Here we only touch upon the spin dynamics, which is responsible for the spin waves and NMR.

4.2. *Soft Modes Dynamics and Lie Algebra of Group G:*
   Spin Dynamics and $SO_3^{(S)}$ Symmetry

Propagating spin waves in ^{3}He-A are coupled oscillations of the spin density $\vec{S}$ with the spin part $\hat{d}$ of the degeneracy parameters. The dynamics of these variables are governed by the Hamiltonian equations, derived by Leggett, which in the A-phase correspond to equations for antiferromagnetic resonance in collinear antiferromagnets.

It is important that the dynamics of the soft modes in condensed matter is also related to the symmetry group G, more exactly with the Lie algebra of this group. In the case of spin dynamics this is the symmetry $SO_3^{(S)}$ subgroup which is relevant, since both the Goldstone field $\hat{d}$ and the spin density $\vec{S}$ change under the action of the spin rotation group $SO_3^{(S)}$. The algebra of the generators $\mathbf{S}_\alpha$ of this group defines the algebra of the Poisson brackets for the hydrodynamic variables in the Hamiltonian approach. To obtain a set of Poisson brackets one must take into account that the hydrodynamical soft variable, spin density $\vec{S}$, is the classical limit of the generator $\mathbf{S}$ of the $SO_3(S)$ group and the Poisson brackets are the classical limit of the commutators, such as $[\mathbf{S}_\alpha, \mathbf{S}_\beta] = i\hbar e_{\alpha\beta\gamma}\mathbf{S}_\gamma$. So one has the following nonzero Poisson brackets, which couple dynamically the components of $\vec{S}$ and those of the degeneracy parameter $\hat{d}$:

$$\{S_\alpha(\vec{r}), S_\beta(\vec{r}')\} = e_{\alpha\beta\gamma} S_\gamma(\vec{r}) \delta(\vec{r} - \vec{r}') , \tag{4.1a}$$

$$\left\{ S_\alpha(\vec{r}), \hat{d}_\beta(\vec{r}') \right\} = e_{\alpha\beta\gamma} \hat{d}_\gamma(\vec{r}) \delta(\vec{r} - \vec{r}') , \tag{4.1b}$$

while the Poisson brackets between the components of $\hat{d}$ are zero.

The Leggett equations in ^{3}He-A are the Liouville equations for $\vec{S}$ and $\hat{d}$:

$$\frac{\partial \vec{S}}{\partial t} = \{\vec{S}, H\} , \tag{4.2a}$$

$$\frac{\partial \hat{d}}{\partial t} = \{\hat{d}, H\} , \tag{4.2b}$$

where the Hamiltonian function H is the energy in terms of the hydrody-namical variables $\vec{S}$ and $\hat{d}$. So the only information we need to deduce the dynamic equations for the given condensed matter is contained in the ex-plicit expression of the energy in terms of the hydrodynamical variables. If one knows H, the Liouville equations become a system of closed hydrody-namical equations:

$$\frac{\partial S_\alpha(\vec{r})}{\partial t} = \int d^3r' \frac{\delta H}{\delta S_\beta(\vec{r}')} \{S_\alpha(\vec{r}), S_\beta(\vec{r}')\} + \int d^3r' \frac{\delta H}{\delta \hat{d}_\beta(\vec{r}')} \{S_\alpha(\vec{r}), \hat{d}_\beta(\vec{r}')\} ,$$
$$\tag{4.2c}$$

$$\frac{\partial \hat{d}_\alpha(\vec{r})}{\partial t} = \int d^3r' \frac{\delta H}{\delta S_\beta(\vec{r}')} \{\hat{d}_\alpha(\vec{r}), S_\beta(\vec{r}')\} + \int d^3r' \frac{\delta H}{\delta \hat{d}_\beta(\vec{r}')} \{\hat{d}_\alpha(\vec{r}), \hat{d}_\beta(\vec{r}')\} .$$
$$\tag{4.2d}$$

The energy H includes i) contributions from the Goldstone fields, which are the London energy of the distortion of the vector $\hat{d}$ in Eq. (3.9a) and spin-orbital energy in Eq. (2.18); and ii) energy of the magnetization $\gamma \vec{S}$ of the liquid, where γ here is the gyromagnetic ratio for the ^{3}He-A nucleus. This is just the magnetic energy (2.14) expressed in terms of the spin density $\vec{S}$. So one has:

$$H = \frac{1}{2}\gamma^2 S_\alpha (\chi^{-1})_{\alpha\beta} S_\beta - \gamma \vec{H} \cdot \vec{S} + F_{\text{grad}}^{\text{London}} + F_{\text{so}} , \tag{4.3}$$

where the anisotropic susceptibility $\chi_{\alpha\beta}$ is given by Eq. (2.13).

Substituting Eq. (4.3) into Eqs. (4.2) and employing the Poisson brackets (4.1), one obtains the closed system of the Leggett equations for the A-phase spin dynamics:

$$\frac{\partial \vec{S}}{\partial t} = \gamma \vec{S} \times \vec{H} - \hat{d} \times \frac{\partial F_{so}}{\partial \hat{d}} + \hat{d} \times \nabla_i \frac{\partial F_{grad}^{London}}{\partial \nabla_i \vec{d}} , \qquad (4.4a)$$

$$\frac{\partial \hat{d}}{\partial t} = \gamma \hat{d} \times \left(\vec{H} - \frac{\gamma \vec{S}}{\chi_\perp} \right) . \qquad (4.4b)$$

4.3. Spin waves. Goldstone and quasi-Goldstone modes

In zero external field, $\vec{H} = 0$, the linearized equations (4.4) for small oscillations of $\hat{d} = \hat{d}_0 + \delta\hat{d}$ and $\vec{S} = \vec{S}_0 + \delta\vec{S}$ about their equilibrium values $\hat{d}_0 = \hat{l}$, $\vec{S}_0 = 0$ lead to the wave equation for $\delta\hat{d}$:

$$\frac{\partial^2 \delta\hat{d}}{\partial t^2} = -\Omega_A^2 \delta\hat{d} + c_\perp^2 \vec{\nabla}^2 \delta\hat{d} + (c_\parallel^2 - c_\perp^2)(\hat{l} \cdot \vec{\nabla})^2 \delta\hat{d} , \qquad (4.5a)$$

which gives the following spectrum for the spin waves:

$$\omega_{1,2}^2(\vec{q}) = \Omega_A^2 + c_\perp^2 q^2 + (c_\parallel^2 - c_\perp^2)(\vec{q} \cdot \hat{l})^2 . \qquad (4.5b)$$

Here the spin-wave velocities are

$$c_{\perp(\parallel)}^2 = \rho_{sp}^{\perp(\parallel)} \gamma^2 / \chi_\perp . \qquad (4.6)$$

The spin wave spectrum contains two degenerate branches which correspond to two degrees of freedom for the deviations of the unit vector $\hat{d}$ from its equilibrium value: $\delta\hat{d} \perp \hat{d}_0$.

Note that the spin waves become non-Goldstone modes in the presence of spin-orbit coupling, since there appears a gap Ω_A in their spectrum:

$$\Omega_A^2 = g_{so}\gamma^2/\chi_\perp = c_\perp^2/\xi_{d\perp}^2 , \qquad (4.7)$$

known as Leggett frequency for the A-phase. Without this coupling the spin waves are true Goldstone bosons: according to the spontaneously broken

$SO_3^{(S)}$ symmetry, their frequency $\omega(\vec{q}) \to 0$ when the wave vector $\vec{q} \to 0$, so no gap is present. However due to the small spin-orbit interaction this $SO_3^{(S)}$ symmetry is no more exact, so the mode acquires a small gap and becomes quasi-Goldstone. This is a general situation: the Goldstone boson always acquires a gap (a mass), if the broken symmetry which leads to the massless boson is approximate or is externally violated.

4.4. Nuclear Magnetic Resonance in ³He-A

The NMR technique provides one of the most versatile tools for extracting information on the structure of the order parameter in the superfluid A and B phases of liquid ³He and on the textures, solitons and vortices in these phases. In the NMR experiments on these ordered liquid magnets, the collective magnetic modes – spin waves – are excited by an external *rf* magnetic field in the presence of the constant magnetic field $\vec{H}_0$. This magnetic field externally violates the $SO_3^{(S)}$ symmetry, which leads to splitting of the degenerate branches in the spin-wave spectrum in Eq. (4.5).

For the equilibrium homogeneous state with $\hat{d}_0 = \hat{l} \perp \vec{H}_0$ the spectrum modifies in the following manner:

$$\omega_{\text{longitudinal}}^2(\vec{q}) = \Omega_A^2 + c_\perp^2 q^2 + (c_\parallel^2 - c_\perp^2)(\vec{q} \cdot \hat{l})^2 , \qquad (4.8)$$

$$\omega_{\text{transverse}}^2(\vec{q}) = \gamma^2 H_0^2 + \Omega_A^2 + c_\perp^2 q^2 + (c_\parallel^2 - c_\perp^2)(\vec{q} \cdot \hat{l})^2 . \qquad (4.9)$$

In the first branch the magnetization oscillates along its equilibrium value $\vec{S}_0 = \chi_\perp \vec{H}_0/\gamma$, i.e. $\delta\vec{S} \parallel \vec{S}_0$ (Fig. 4.1a), and these oscillations are excited by the longitudinal *rf* field, $\vec{H}_{rf} \parallel \vec{H}_0$. The $\hat{d}$ vector oscillates in the transverse plane, which corresponds to the spin rotations about the field direction. The magnetic field does not violate the symmetry under such rotations, therefore the longitudinal oscillations still represent the Goldstone mode: there is no gap in Eq. (4.8) if one neglects the gap Ω_A due to dipole forces. The corresponding resonance absorption of the *rf* signal on the spin waves is called longitudinal NMR.

In the second branch the magnetization precesses around its equilibrium value $\vec{S}_0 = \chi_\perp \vec{H}_0/\gamma$, with $\delta\vec{S} \perp \vec{S}_0$, and these small oscillations are excited by the transverse *rf* field , $\vec{H}_{rf} \perp \vec{H}_0$. The corresponding resonance absorption of the *rf* signal on the transverse spin waves is called transverse NMR.

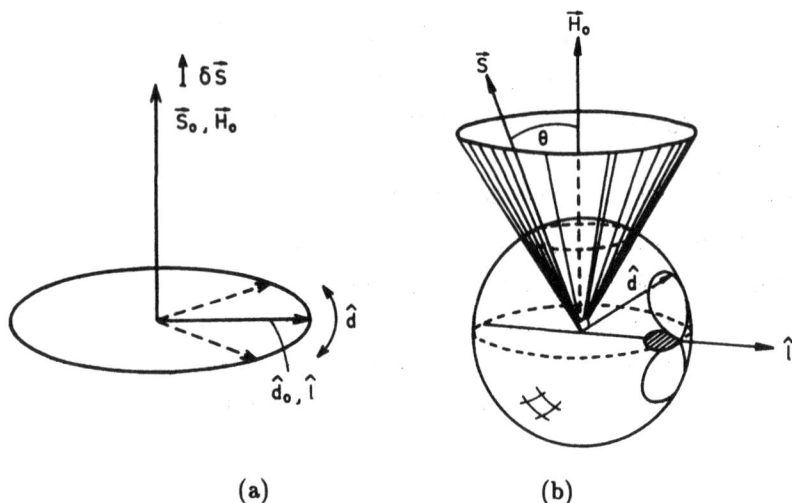

Fig. 4.1. Oscillations of the magnetization $\vec{S}$ and of the degeneracy parameter $\hat{d}$ in the longitudinal (a) and transverse (b) NMR.

For this branch of spectrum the $\hat{d}$ vector is not confined in the transverse plane (Fig. 4.1b), so other $SO_3^{(S)}$ rotations are involved. As a result an additional gap appears in the Goldstone mode. It is produced by the magnetic field which externally violates the $SO_3^{(S)}$ rotational symmetry.

In the general case of the inhomogeneous order parameter the spectrum of spin waves depends on the mutual orientation of the vectors $\hat{d}$, $\hat{l}$ and $\vec{H}_0$ in an essential manner; this gives the possibility to detect different order parameter textures.

4.5. *NMR on Textures in ^3He-A*

In NMR experiments on superfluid ^3He textures, mostly the transverse NMR is used. In this case the rf field $\vec{H}_{rf}$ is transverse to the constant magnetic field $\vec{H}_0 = H_0\hat{z}$, which is chosen to be large enough as compared with the effective field $H_d \sim$ 20–50 Gauss, produced by spin-orbital coupling. In such a field, the equilibrium $\hat{d}$ texture produced by competing orientating effects of $\vec{H}$ and $\hat{l}$ is locked in the transverse $(\hat{x}\hat{y})$-plane:

$$\vec{d}_0(\vec{r}) = \hat{x}\cos\alpha + \hat{y}\sin\alpha \ .$$

The spin modes excited in the transverse NMR are coupled oscillations of the transverse components of the spin density $\delta\vec{S} \perp \vec{S}_0$ and the longitudinal (along $\vec{H}_0$) deviation of $\vec{d}$ from its equilibrium value, thus $\hat{d}(\vec{r}, t) = \hat{d}_0(\vec{r}) + \psi(\vec{r}, t)\hat{z}$ with $\psi \ll 1$.

Using Eq. (4.4b), one may express the $\hat{z} \times \hat{d}_0$ component of spin density $\delta\vec{S}$ through ψ:

$$(\hat{z} \times \hat{d}_0) \cdot \delta\vec{S} = -i\frac{\partial\psi}{\partial t}\frac{\chi_\perp}{\gamma^2} \ , \tag{4.10}$$

and from Eq. (4.4a), with scalar multiplication by $\hat{d}$, the other relevant component of $\delta\vec{S}$ which is parallel to $\vec{d}_0$ is obtained in terms of ψ:

$$\hat{d}_0 \cdot \vec{S}_\perp = -\psi H_0\frac{\chi_\perp}{\gamma} \ . \tag{4.11}$$

Finally, after scalar multiplication of Eq. (4.4a) by $\hat{z} \times \hat{d}_0$, and expressing $\delta\vec{S}$ through ψ by means of Eqs. (4.10) and (4.11), we obtain the wave equation obeyed by the transverse spin mode:

$$-\frac{\partial^2\psi}{\partial t^2} = \left(\gamma^2 H_0^2 + \Omega_A^2\right)\psi + \Omega_A^2\left(U(\vec{r})\psi + \mathbf{D}\psi\right) \ , \tag{4.12}$$

where the spin-wave potential $U(\vec{r})$ for transverse spin modes is given by

$$U(\vec{r}) = -\left[(\hat{l} \times \hat{d}_0)^2 + \frac{(\hat{l} \cdot \vec{H}_0)^2}{H_0^2}\right] + \xi_{d\perp}^2(\nabla_i\hat{d})^2 - (\xi_{d\|}^2 + \xi_{d\perp}^2)\left((\hat{l}\cdot\vec{\nabla})\hat{d}\right)^2 \ , \tag{4.13}$$

and $\mathbf{D}$ is the kinetic-energy operator

$$\mathbf{D}\psi = -\xi_{d\perp}^2\vec{\nabla}^2\psi - (\xi_{d\|}^2 - \xi_{d\perp}^2)\vec{\nabla}\cdot\left(\hat{l}(\hat{l}\cdot\vec{\nabla})\psi\right) \ . \tag{4.14}$$

In the case of a homogeneous liquid with $\hat{l} \parallel \hat{d}_0 \perp \vec{H}_0$ and $U = 0$ the transverse NMR frequency is the frequency of excited spin waves, whose spectrum as found from Eq. (4.12) gives Eq. (4.9).

Textures produce the potential Eq. (4.13) in the wave equation, which becomes attractive if $\hat{l}$, rather than $\hat{d}_0$, changes in space. This takes place in the soft cores of continuous vortices and in some types of solitons (for the pure $\hat{d}$ soliton considered in Sec. 3.5 with only the $\hat{d}$ vector changing in space, the potential is exactly zero, which is a consequence of the translational symmetry of the $\hat{l}$ field). These solitons and vortices produce potential wells which are one-dimensional and two-dimensional respectively. In both wells, there always exists a bound state with the localized spin-wave frequency below the continuum part of the spectrum in Eq. (4.9):

$$\omega_{\text{loc}}^2 = \gamma^2 H_0^2 + \Omega_A^2 + E\Omega_A^2 \ , \quad E < 0 \ ; \tag{4.15}$$

where E is the eigenvalue of the "Schrödinger" equation

$$E\psi = \mathbf{D}\psi + U(\vec{r})\psi \ . \tag{4.16}$$

The excitation of localized spin modes results in an additional (satellite) absorption peak at the frequency ω_{loc} in the NMR experiments on solitons and vortices. The experimental magnitudes of ω_{loc} give some information on vortex and soliton structures.

4.6. *Vacuum Symmetry H and Quantum Numbers of Bose and Fermi Excitations in* ^3He-A

In particle physics the elementary particles are classified in terms of quantum numbers: spin, electric charge, baryon number, parity, isospin, chirality, etc.... . All these quantum numbers result from the symmetry of the vacuum, for example the spin quantum number is the consequence of the isotropy of the vacuum state, i.e. of the SO_3 symmetry, and the charge quantization comes from the $U(1)$ gauge symmetry.

In the same way quasiparticles in condensed matter, such as fermions and bosonic collective modes in ^{3}He-A, have quantum numbers defined by the symmetry H of the equilibrium or vacuum state of a given system. The total group H of the equilibrium ^{3}He-A state contains two continuous Abelian groups: i) Spin rotations $SO_2^{(S)}$ about the axis $\hat{d}$, and ii) the combined continuous symmetry $U(1)^{\text{combined}}$: the vacuum state is invariant under the

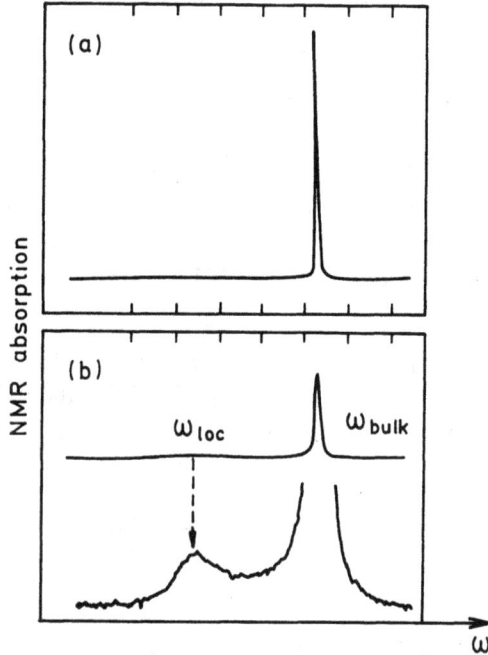

Fig. 4.2. The quantized vortices in rotating ^{3}He-A were detected due to excitations of the spin-wave modes, localized in the soft cores of continuous vortices. In the nonrotating state (a) the conventional absorption at the frequency $\omega_{bulk} = \sqrt{\gamma^2 H_0^2 + \Omega_A^2}$ of the transverse NMR is shown. Under rotation (b) the localized modes produce an additional satellite peak in the NMR absorption at frequency ω_{loc} below ω_{bulk} of the transverse NMR on the bulk A-phase. Two different scales of absorption are shown.

gauge transformation from $U(1)$ group, if it is simultaneously accompanied by the orbital rotation θ_3 about axis $\hat{l}$.

There are also three discrete symmetries, which we denote as P_1, P_2 and P_3. iii) The P_1 group is combined symmetry $Z_2^{combined}$, which was discussed in Secs. 2.9 and 2.11.

iv) The P_2 symmetry transformation is the combined transformation $P_2 = TC_{\pi,x}^L$. Here T is the time inversion symmetry which corresponds to the complex conjugation for the order parameter:

$$\mathbf{T}(\hat{e}^{(1)} + i\hat{e}^{(2)}) = \hat{e}^{(1)} - i\hat{e}^{(2)} \ , \ \mathbf{T}A_{\alpha i} = A_{\alpha i}^* \ . \tag{4.17}$$

The time inversion thus changes the sign of $\hat{e}^{(2)}$. The symmetry operation $C_{\pi,x}^L$ is the π rotation of orbital space about axis $\hat{e}^{(1)}$. This transformation also changes the sign of $\hat{e}^{(2)}$ and thus restores the initial value of the degeneracy parameter $\hat{e}^{(1)} + i\hat{e}^{(2)}$, if both transformations are applied simultaneously.

And finally v) the P_3 group is related to the conventional space parity. In superfluid ³He the spatial parity P is also broken, since under this transformation the order parameter changes sign:

$$\mathbf{P}A_{\alpha i} = -A_{\alpha i} \ . \tag{4.18}$$

This comes from the fact that the parity of the Cooper pair is $P = (-1)^L$ with $L = 1$ for the superfluid ³He. However the combined symmetry exists, which we denote by $P_3 = P \cdot e^{i\pi}$. This is the combination of parity with the gauge transformation by the phase π. For all those physical quantities which are gauge invariant, this combined symmetry is the conventional space inversion symmetry.

The discrete symmetry, like parity, gives rise to a quantum number with only two values $+$ and $-$. The continuous Abelian symmetry of the $U(1)$ type leads to a quantum number which can take any integer value. This number is the eigenvalue of the generator of the group.

For the group $SO_2^{(S)}$ the generator is the projection $\mathbf{S}_z$ of the spin momentum operator $\mathbf{S}$ on the $\hat{d}$ axis, whose eigenvalues M^S are integers for bosons and half-integers for fermions. The action of the spin momentum operator $\mathbf{S}$ on the order parameter matrix $A_{\alpha i}$ is

$$\mathbf{S}_\beta A_{\alpha i} = -i e_{\beta\alpha\gamma} A_{\gamma i} \ . \tag{4.19}$$

For another continuous symmetry which is the combination of $U(1)$ and

$SO_2^{(L)}$, the generator is a linear combination of the generators of the corresponding group

$$Q = \frac{1}{2}I - L_z \,, \qquad (4.20)$$

whose eigenvalue we denote by Q. The operator of the orbital rotations acts on the order parameter matrix as follows:

$$L_k A_{\alpha i} = -i e_{kil} A_{\alpha l} \,. \qquad (4.21)$$

And the generator I of the gauge transformation is the operator of the particle number for the ^{3}He atoms; it has the value 2 for the Cooper pair order parameter, and -2 for its complex conjugation (here we take into account that the Cooper pair consists of two particles):

$$I A_{\alpha i} = 2 A_{\alpha i} \,, I A_{\alpha i}^* = -2 A_{\alpha i}^* \,. \qquad (4.22)$$

So in general each Bose or Fermi excitation in superfluid ^{3}He-A is described by 5 quantum numbers Q, S_z, P_1, P_2, P_3.

4.7. Bosonic Collective Modes in ^{3}He-A and Irreducible Representations of Group H

Of the collective bosonic modes, 18 modes are especially distinguished. These are the oscillations of the 18 components of the deviations of the order parameter $A_{\alpha i}$ from its equilibrium value:

$$A_{\alpha i} = A_{\alpha i}^0 + \delta A_{\alpha i} = \Delta_A(T)\left(\hat{z}_\alpha(\hat{x}_i + i\hat{y}_i) + u_{\alpha i} + i v_{\alpha i}\right) \,, \qquad (4.23)$$

where $u_{\alpha i}$ and $v_{\alpha i}$ are (small) real dimensionless variables. Using Eqs. (4.17)–(4.22) for the symmetry operations one can find the quantum numbers for these 18 modes of the order parameter oscillations (all with positive $P_3 = +$):

$\lvert Q \rvert$	$\lvert M_S \rvert$	D	P_2	P_1	Variables	Modes
0	0	1	−	+	$u_{zy} - v_{zx}$	Sound (oscillations of $\mathbf{v}_s$)
0	1	2	+	+	$u_{yx} + v_{yy}$	Spin waves (Higgs bosons)
			+	−	$u_{xx} + v_{xy}$	(oscillations of $\hat{d}$)
1	0	2	+	+	v_{zz}	Orbital waves (photons)
			−	+	u_{zz}	(oscillations of $\hat{l}$)
1	1	4	+	+	v_{yz}	Spin-orbital waves(W bosons)
			+	−	v_{xz}	
			−	+	u_{yz}	
			−	−	u_{xz}	
0	0	1	+	+	$u_{zx} + v_{zy}$	Pair-breaking mode
0	1	2	−	−	$u_{xy} - v_{xx}$	Pair-breaking modes
			−	+	$u_{yy} - v_{yx}$	
2	0	2	+	+	$u_{zx} - v_{zy}$	Clapping modes (gravitons)
			−	+	$u_{zy} + v_{zx}$	
2	1	4	+	+	$u_{yx} - v_{yy}$	Clapping modes
			−	+	$u_{yy} + v_{yx}$	
			+	−	$u_{xx} - v_{xy}$	
			−	−	$u_{xy} + v_{xx}$	

$$(4.24)$$

In the last column the analogues of the bosons in particle physics are given in brackets. In the third column, D means degeneracy of the collective modes (or the dimension of the representation of the group H). The degeneracy results from the fact that not all elements of the group H commute with one another. For example, the elements of spin rotations $SO_2^{(S)}$ about z axis do not commute with P_1 since the latter contains rotation about the transverse axis. This leads to the double degeneracy of the modes with $M_S \neq 0$. So if the mode has definite parity P_1, it has no well-defined M_S since it is the combination of states with $M_S = +1$ and $M_S = -1$. On the contrary, states with well-defined $M_S = +1$ and $M_S = -1$ transform into each other under the parity P_1 transformation, forming thus the two-dimensional representation of the group H.

In the same manner, Q does not commute with P_2, leading to double degeneracy of the modes with $Q \neq 0$. As a result the 18-dimensional representation of the group G for the collective modes of the order parameter oscillations $\delta A_{\alpha i}$ is reduced to 8 irreducible representations of the group H: 2 one-dimensional, 4 two-dimensional and 2 with D=4. So the collective modes in a given superfluid phase are classified in terms of irreducible representations of the residual symmetry H of the phase.

The collective modes of the oscillations of the principal order parameter $A_{\alpha i}$ are observed in ultrasonic and NMR experiments. The collective modes with other quantum numbers are also possible, they correspond to excitation of the Cooper pair states with L and S different from 1. However there is still no solid experimental evidence of their existence. For example, while the amplitude ψ of the conventional Cooper pairing with $L = 0$ and $S = 0$ is exactly zero in the equilibrium state of the A phase, the oscillations of ψ on the background of the A-phase vacuum are in principle possible. As we shall see in Sec. 5 these doubly degenerate collective modes (ψ is complex amplitude), with $|Q| = 1$, $P_2 = +$ for the real part of ψ and $P_2 = -$ for the imaginary part, $M_S = 0$, $P_3 = -$, $P_1 = -$, correspond to the Z intermediate bosons in particle physics.

4.8. Bosonic Collective Modes in ^{3}He-B

In the same way, 18 modes of the order parameter oscillations are distinguished in the B-phase:

$$A_{\alpha i} = A^0_{\alpha i} + \delta A_{\alpha i} = \Delta_B(T) R_{\alpha k}(\delta_{ki} + u_{ki} + iv_{ki}) , \qquad (4.25)$$

which should be distributed into irreducible representations of the residual symmetry group H for the B-phase:

$$H = SO_3^{(J)} \times T \times P_3 . \qquad (4.26)$$

The representations of this group are classified by the quantum numbers
i) $T = +$ or $-$, ii) $P_3 = +$ for all 18 modes, and iii) the magnitude $J = 0$,
1 , 2 of the total angular momentum **J**. Each mode has $2J + 1$ degeneracy,
which is lifted by the magnetic field (see Fig. 4.3).

J	T	D	Variables	Modes
0	$-$	1	v_{ii}	Sound
0	$+$	1	u_{ii}	Pair-breaking mode
1	$-$	3	$v_{ik} - v_{ki}$	Pair-breaking modes
1	$+$	3	$u_{ik} - u_{ki}$	Spin waves
2	$-$	5	$v_{ik} + v_{ki} - \frac{2}{3}\delta_{ik}v_{ll}$	Squashing modes
2	$+$	5	$u_{ik} + u_{ki} - \frac{2}{3}\delta_{ik}u_{ll}$	Real squashing modes (gravitons)

$$(4.27)$$

The collective modes with higher $J > 2$ are also possible, they correspond
to oscillations of the Cooper pair amplitudes with higher $L > 1$, which are
exactly zero in the unperturbed equilibrium state of the B-phase but can be
excited dynamically.

4.9. Dynamics of the Goldstone Fields in ³He-B

Four of the 18 modes in Eq. (4.27) are Goldstone bosons: i) 3 polarizations of the spin waves ($J = 1$, $M_J = -1, 0, +1$), which are coupled
oscillations of the degeneracy parameter matrix $R_{\alpha i}$ and the spin density
$\vec{S}$; and ii) sound which is the coupled oscillation of the phase Φ of the Bose
condensate and the mass (or particle) density ρ.

The soft modes dynamics is regulated by the Poisson brackets resulting
from the Lie algebra (commutation relations) for the generators **S** and **I** of
the $SO_3^{(S)} \times U(1)$ group:

$$\{S_\alpha(\vec{r}), S_\beta(\vec{r}')\} = e_{\alpha\beta\gamma}S_\gamma(\vec{r})\delta(\vec{r} - \vec{r}') , \qquad (4.28a)$$

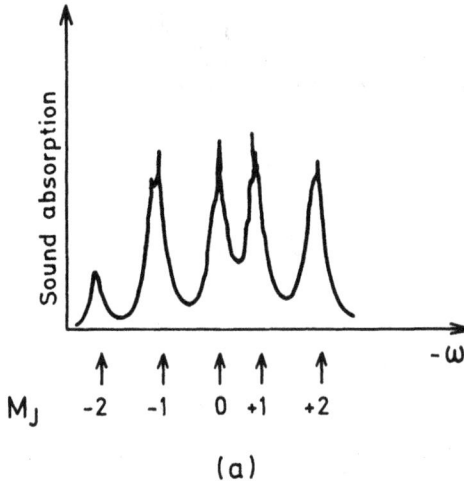

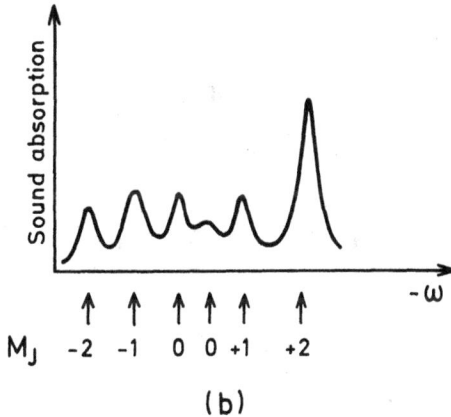

Fig. 4.3. (a) 5-fold splitting of the real squashing mode ($J = 2, T = +$) in magnetic field. (b) Additional splitting of the mode with projection $M_J = 0$ arising from the textures. This is the result of the broken relative spin-orbital symmetry: the actual axis of the quantization of M_J is not the direction of the magnetic field, but the direction obtained by the order parameter rotation: $R_{\alpha i} H_\alpha$. In the presence of textures in the $R_{\alpha i}$ field the quantization axis becomes coordinate-dependent, which is observed as an additional splitting.

$$\left\{S_\alpha(\vec{r}), R_{\beta i}(\vec{r}\,')\right\} = e_{\alpha\beta\gamma}R_{\gamma i}(\vec{r})\delta(\vec{r} - \vec{r}\,') \,, \tag{4.28b}$$

$$\left\{\rho(\vec{r}\,), \Phi(\vec{r}\,')\right\} = M\delta(\vec{r} - \vec{r}\,') \,. \tag{4.29}$$

The Liouville equations for $\vec{S}$, $R_{\alpha i}$, ρ and Φ are

$$\frac{\partial \vec{S}}{\partial t} = \{\vec{S}, H\} \,, \tag{4.30a}$$

$$\frac{\partial R_{\alpha i}}{\partial t} = \{R_{\alpha i}, H\} \,, \tag{4.30b}$$

$$\frac{\partial \rho}{\partial t} = \{\rho, H\} \,, \tag{4.31a}$$

$$\frac{\partial \Phi}{\partial t} = \{\Phi, H\} \,. \tag{4.31b}$$

The energy H, if one neglects the spin-orbit coupling, is

$$H = \epsilon(\rho) + \frac{1}{2}\frac{\gamma^2}{\chi}S^2 - \gamma\vec{H}\cdot\vec{S} + F_{\text{grad}}^{\text{London}} \,, \tag{4.32}$$

where the susceptibility χ is isotropic in the B-phase, $\epsilon(\rho)$ is the density dependence of the energy of the liquid, and the London energy of the B-phase is:

$$F_{\text{grad}}^{\text{London}} = \frac{1}{2}\rho_s\vec{v}_s^2 + \frac{1}{2}K_1(\nabla_i R_{\alpha k})^2 + \frac{1}{2}K_2(\nabla_i R_{\alpha i})^2 \,. \tag{4.33}$$

4.10. *Superfluid Hydrodynamics in* ^3He-*B*

Using the Hamiltonian (4.32) one obtains from Eq. (4.31) the equations for the superfluid hydrodynamics, which corresponds to that of superfluid ^4He:

$$\partial_t\rho + \vec{\nabla}\cdot\vec{j} = 0 \,, \tag{4.34}$$

$$\partial_t\Phi = -\frac{2m_3}{\hbar}\mu \,, \quad \text{or} \quad \partial_t\vec{v}_s = -\vec{\nabla}\mu \,. \tag{4.35}$$

Here $\vec{j} = \rho_s\vec{v}_s$ is the mass supercurrent with the superfluid density corresponding to the total mass density at zero temperature, and the chemical potential $\mu = \frac{\delta E}{\delta \rho}$.

The second equation is characteristic of superfluid hydrodynamics of conventional superfluids with curl-free superfluid velocity. (For the A-phase where the superflow is not irrotational the equation for the superfluid velocity is different, see Sec. 6). It shows that within the applicability of this equation the superfluid velocity increases without limit under external force ($\vec{\nabla}\mu$ may be created by the temperature or pressure gradients or by electric field for an electrically charged system). This process principally cannot be stopped by any conventional friction forces which are so effective in the ordinary viscous liquid leading to the relaxation of motion due to momentum transfer from the liquid to the walls of the container. As we know, in the superfluid liquid the uniform superflow is dissipationless since this corresponds to the texture; therefore there is no friction force which is proportional to the difference between $\vec{v}_s$ and the velocity of the container wall or proportional to $\vec{v}_s - \vec{v}_n$, where $\vec{v}_n$ is the velocity of the normal component of liquid at nonzero T.

The process, which limits the catastrophical increase of superfluid velocity in conventional superfluids, is the textural dynamics of the phase Φ which starts from some critical velocity. The increase of the phase difference is compensated by abrupt 2π (singular) jumps of the phase, the phase slip, when, for example, the singularities in the order parameter field, the quantized vortices, cross the channel. In the steady-state supercritical flow the phase slip processes compensate on average the gradient of the chemical potential, so the average $< \vec{v}_s >=$ const. The processes of the phase slippage occur, as a rule, periodically, resulting in the ac Josephson effect: periodic oscillations under an external driving force constant in time.

4.11. *Goldstone Bosons in* ^{3}*He-B*

In the linearized version the Liouville equations (4.30), (4.31) are a system of four wave equations for the phase Φ and for small angle $\vec{\theta}$ of the solid rotations of the matrix $R_{\alpha i}$:

$$\delta R_{\alpha i} = e_{ijk}\theta_j R_{\alpha k} , \quad \theta_i = \frac{1}{2}e_{ijk}u_{jk} . \qquad (4.36)$$

These equations are

$$\frac{\partial^2 \Phi}{\partial t^2} = c_4^2 \nabla^2 \Phi , \qquad (4.37)$$

$$\frac{\partial^2 \vec{\theta}}{\partial t^2} = c_1^2 \vec{\nabla}^2 \vec{\theta} + (c_2^2 - c_1^2) \vec{\nabla}(\vec{\nabla} \cdot \vec{\theta}) . \tag{4.38}$$

Here c_4 is the fourth sound velocity $c_4^2 = \rho_s \frac{\partial^2 \epsilon}{\partial \rho^2}$. The fourth sound propagates under such conditions that the motion of the normal component is suppressed, say, by the wall of the container. c_1 is the spin wave velocity for the waves with polarization $\vec{\theta} \perp \vec{q}$ ($c_1^2 = 2K_1\gamma^2/\chi$), and c_2, with $c_2^2 = (2K_1 + K_2)\gamma^2/\chi$, is the velocity of the spin waves with polarization $\vec{\theta} \parallel \vec{q}$.

5

Fermi Excitations in Superfluid Phases of ³He

5.1. Bogoliubov-Nambu Matrix for the Fermions in Pair-Correlated Fermi Systems

The quasiparticle energy spectrum $E_{\vec{k}}$ in a Fermi system with Cooper pairing is determined by the eigenvalues of the Bogoliubov matrix

$$\mathbf{H} = \begin{pmatrix} \varepsilon_{\vec{k}} & \Delta(\vec{k}) \\ \Delta^{\dagger}(\vec{k}) & -\varepsilon_{\vec{k}} \end{pmatrix} . \tag{5.1}$$

Here $\varepsilon_{\vec{k}} = \epsilon(\vec{k}) - \mu$ is the quasiparticle energy in the normal Fermi liquid, counted from the chemical potential μ. In the normal Fermi liquid the chemical potential is positive and $\varepsilon_{\vec{k}}$ turns to zero on the whole surface $k = k_F$, where $\epsilon(k_F) = \mu$. This is the Fermi surface of the normal Fermi liquid; in the vicinity of the Fermi surface $\varepsilon_{\vec{k}} \approx v_F(k - k_F)$, where v_F and k_F are the Fermi velocity and Fermi momentum respectively. In the limiting case of the ideal Fermi gas, $\varepsilon_{\vec{k}} = (k^2 - k_F^2)/2m_3$.

The phenomenon of Cooper pairing, at which the Cooper pairs of ³He atoms form a coherent Bose condensate, displays itself in the appearance of the nonzero matrix element $\Delta(\vec{k})$ between the particle state with energy $\varepsilon_{\vec{k}}$ and the hole state with the opposite energy $-\varepsilon_{\vec{k}}$. In the pair condensate the states which differ by an even number of particles (say N and $N \pm 2$)

are indistinguishable, therefore the particle-hole transmutation is possible through interaction with the condensate, since these states just differ by two ³He atoms. Usually $\Delta(\vec{k}) \ll \epsilon(k_F)$, so the Cooper pairing while producing the off-diagonal elements does not influence much the diagonal terms in the Bogoliubov matrix.

For the particles with spin 1/2 all the elements of the Bogoliubov matrix are also 2×2 spin matrices: $\varepsilon_{\vec{k}} \to \varepsilon_{\vec{k}} \delta_b^a$ and $\Delta(\vec{k}) \to \Delta(\vec{k}) = \Delta_b^a(\vec{k})$, where a and b are spin indices which take two values (↑ and ↓), so **H** is a 4×4 matrix. The off-diagonal element $\Delta_b^a(\vec{k})$, which appears below T_c, may be treated as the order parameter, and it transforms according to the relevant representation of the symmetry group $G = U(1) \times SO_3^{(S)} \times SO_3^{(L)}$:

$$\Delta(\vec{k}) \to e^{2i\alpha} \mathbf{U} \Delta(R_{ij}^{(L)} k_j) \mathbf{U}^\dagger , \qquad (5.2)$$

where **U** is the unitary matrix of the spin rotation group $SO_3^{(S)}$, while the orbital group $SO_3^{(L)}$ rotates the momentum $\vec{k}$. Under the gauge transformation $\Delta(\vec{k})$ is multiplied by the phase factor $e^{2i\alpha}$ which contains the factor 2 since the off-diagonal matrix element couples the states which differ by two particles.

5.2. *Quasiparticles in Conventional Superconductors*

In the case of conventional superconductors one usually exploits a model in which the crystal field is neglected and the Fermi liquid of the electrons in the normal metal state is considered to be isotropic. In this case the normal metal state has also the symmetry $G = U(1) \times SO_3^{(S)} \times SO_3^{(L)}$ if the spin orbital coupling is neglected. The relevant representation of this group for conventional superconductors has quantum numbers $S = 0$, $L = 0$, i.e. the superconducting state conserves all the symmetry of the normal state excluding the gauge symmetry $U(1)$. So the gap function is a scalar function which is isotropic in momentum and spin space: $\Delta_b^a(\vec{k}) = \delta_b^a \Delta$. The dependence on $|\vec{k}|$ is unimportant since k is concentrated mainly in the vicinity of the Fermi-momentum k_F. The complex scalar Δ, which transforms under the gauge transformation as $\Delta \to \Delta e^{2i\alpha}$, is the standard order parameter in conventional superconductors, often denoted as ψ in the Ginzburg-Landau theory.

The quasiparticle energy spectrum $E_{\vec{k}}$ can be obtained from Eq. (5.1) taking into account that the square of the Hamiltonian $\mathbf{H}^2$ is diagonal in this simple case

$$\mathbf{H}^2 = \begin{pmatrix} E_{\vec{k}}^2 & 0 & 0 & 0 \\ 0 & E_{\vec{k}}^2 & 0 & 0 \\ 0 & 0 & E_{\vec{k}}^2 & 0 \\ 0 & 0 & 0 & E_{\vec{k}}^2 \end{pmatrix} ,$$

with
$$E_{\vec{k}}^2 = v_F^2(k - k_F)^2 + |\Delta|^2 ,$$

so the modulus of Δ gives a gap in the fermionic spectrum, which is thus the same for all directions of $\vec{k}$ in this isotropic case (Fig. 5.1).

If the crystal fields are taken into account, both the initial quasiparticle energy $\varepsilon_{\vec{k}}$ and the gap become dependent on the momentum direction. For conventional superconductors, $\Delta(\vec{k})$ has all the point symmetries of the crystals, i.e., has the same symmetry properties as $\varepsilon_{\vec{k}}$ except the gauge invariance. The gap function may be represented as $\Delta(\vec{k}) = f(\vec{k})\psi$, where $f(\vec{k})$ is a real function with the symmetry of $\varepsilon_{\vec{k}}$ and ψ is the complex order parameter responsible for the broken gauge symmetry. The energy spectrum in this anisotropic case is

$$E_{\vec{k}}^2 = \varepsilon_{\vec{k}}^2 + |\Delta(\vec{k})|^2 ,$$

so the gap depends on the momentum direction.

In the isotropic case, due to the gap in the spectrum, the fermionic excitations are frozen out exponentially, as $\exp(-|\Delta|/T)$, at low enough temperature, $T \ll T_c \sim |\Delta|$. For the real crystal usually there is no reason for the gap function $\Delta(\vec{k})$ to turn to zero at some points, lines or surfaces in momentum space. So the Fermi manifold – the set of points in the momentum space where the quasiparticle spectrum crosses zero energy – is present in normal state in the form of the Fermi surface and completely disappears in conventional superconductors.

As we shall see below this is not the case for the unconventional superfluids or superconductors, where the gauge symmetry is relatively broken

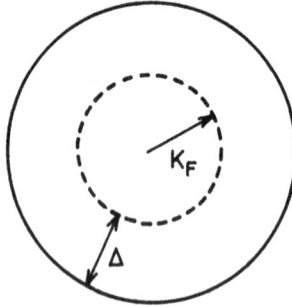

Fig. 5.1. Isotropic gap in the quasiparticle spectrum in the superfluid ³He-B and in the simplest model of isotropic superconductor. Dashed line shows Fermi surface in the normal Fermi liquid – the manifold of points in momentum space where the quasiparticle energy is zero. In the superconducting state of a conventional superconductor the gap in the spectrum leads to complete disappearance of the Fermi manifold.

and the gap function has a lower symmetry than $\varepsilon_{\vec{k}}$. The superfluid ³He-A is an example of the systems where the gap function turns to zero at two points of the momentum space. This also can occur in heavy-fermion superconductors, where points and/or lines of zeroes in the gap are supposed to exist due to additional breaking of the crystal symmetry. So in unconventional superconductors or superfluids the Fermi manifolds in the quasiparticle spectrum may still survive as Fermi points or Fermi lines, or even, sometimes, as Fermi surface.

5.3. Representation of the Gap Function in ³He and the Order Parameter

In superfluid ³He the relevant representation has quantum numbers $S = 1$, $L = 1$, i.e., this is the product of the $S = 1$ representation, given by the 2×2 Pauli matrices σ_x, σ_y, and σ_z, and the $L = 1$ representation, given by the components of the vector $\vec{k}$ in the orbital space. The corresponding amplitudes in the expansion of $\Delta_b^a(\vec{k})$ in terms of the basis functions σ_α and k_i,

$$\Delta_b^a(\vec{k}) = A_{\alpha i} \frac{k_i}{k_F} (\sigma_\alpha)_b^a , \qquad (5.3)$$

form the 3×3 order parameter matrix $A_{\alpha i}$ which we used in previous sections. We can ascribe the transformation properties of $\Delta(\vec{k})$ to this matrix $A_{\alpha i}$, so it transforms as a vector under a spin rotation for a fixed orbital index (i) – and as a vector under an orbital rotation for a fixed spin index (α).

5.4. *Quasiparticle Spectrum in* ³He-B

Introducing an equilibrium order parameter for the B-phase, $A_{\alpha i}^{(0)} = \Delta_B \delta_{\alpha i}$, one obtains the following Bogoliubov matrix:

$$\mathbf{H} = \begin{pmatrix} \varepsilon_{\vec{k}} & \dfrac{\Delta_B}{k_F}\vec{\sigma}\cdot\vec{k} \\ \dfrac{\Delta_B}{k_F}\vec{\sigma}\cdot\vec{k} & -\varepsilon_{\vec{k}} \end{pmatrix} = \tau_3(\epsilon(\vec{k}) - \mu) + \frac{\Delta_B}{k_F}\tau_1\vec{\sigma}\cdot\vec{k} \ , \qquad (5.4)$$

where the Pauli 2×2 matrices $\vec{\tau}$ describe the Bogoliubov isospin in the particle-hole space:

$$\tau_3 = \begin{pmatrix} 1 & 0 \\ 0 & -1 \end{pmatrix} , \quad \tau_1 = \begin{pmatrix} 0 & 1 \\ 1 & 0 \end{pmatrix} , \quad \tau_2 = \begin{pmatrix} 0 & -i \\ i & 0 \end{pmatrix} . \qquad (5.5)$$

The quasiparticle spectrum is obtained again when calculating $\mathbf{H}^2$, which also proves to be diagonal:

$$E_{\vec{k}}^2 = (\epsilon(\vec{k}) - \mu)^2 + \Delta_B^2 \frac{k^2}{k_F^2} \approx v_F^2(k - k_F)^2 + \Delta_B^2 \ . \qquad (5.6)$$

Here we took into account that the gap $\Delta_B \ll \epsilon(k_F)$, and as a result the momenta $\vec{k}$ are concentrated near the old Fermi surface, $k \approx k_F$. So in ³He-B the gap in the quasiparticle spectrum is isotropic, without any nodes as in a conventional superconductor. Nevertheless, the physics of the fermionic quasiparticles in ³He-B is much richer than in a conventional superconductor due to the spin degrees of freedom and more complicated structure of the order parameter, whose components produce external fields for the fermions.

5.5. *Bogoliubov Hamiltonian for Quasiparticles in* ³He-B *vs. Dirac Hamiltonian for Electrons*

The 4×4 matrix H for the ³He-B quasiparticles may be expressed in terms of the Dirac 4×4 matrices $\vec{\alpha}$ and β:

$$\beta = \tau_3 \ , \quad \vec{\alpha} = \tau_1\vec{\sigma} = \begin{pmatrix} 0 & \vec{\sigma} \\ \vec{\sigma} & 0 \end{pmatrix} \qquad (5.7)$$

in the following form which reminds one of the Dirac Hamiltonian for the electrons with mass m:

$$\mathbf{H} = \beta m(\vec{k})c^2 + c\vec{\alpha} \cdot \vec{k} , \qquad (5.8)$$

where the corresponding speed of light c and the mass $m(\vec{k})$ are

$$c = \Delta_B/k_F , \quad m(\vec{k}) = \left(\frac{\vec{k}^2}{2m_3} - \mu\right)/c^2 . \qquad (5.9)$$

As distinct from the Dirac equation for the electron, the mass term essentially depends on k. However there exists one limiting case when the Bogoliubov Hamiltonian exactly coincides with the Dirac one. This is the case of very strong coupling between the ³He atoms when the gap $\Delta_B \gg \epsilon(k_F)$. In this case the Cooper pairing influences also the diagonal terms in the Hamiltonian and the chemical potential μ is no more $\epsilon(k_F)$, but decreases and at some moment becomes negative. After that the low energy excitations will be concentrated near $k = 0$. In this limit the term $k^2/2m_3$ may be neglected for the low-energy (low momentum) excitations, if $m_3c^2 \gg |\mu|$. So at $k \to 0$ the mass term in Eq. (5.8) may be considered as constant, $mc^2 = -\mu > 0$, and the low energy quasiparticle spectrum becomes relativistic, $E^2 = m^2c^4 + c^2k^2$.

Let us consider the symmetry of the Dirac Hamiltonian from the point of view of the ³He-B. What is most important for the Dirac equation is the symmetry under the Lorentz transformations which form the Lorentz group L. This is the group of rotations in the 4-dimensional space-time, which includes the SO_3 group of the space rotations and group of transformations to the moving coordinate frame. In ³He only the first subgroup is retained, since any condensed matter with finite mass density is not invariant under the transformation to the moving frame, so we shall consider here the SO_3 group of space rotations and then generalize the consequences to the case of the total Lorentz group. The analogy between the Bogoliubov and Dirac equations may give some insight into the origin of the symmetry of the Dirac equation: in particular, from the ³He-B point of view, one may suggest that the Lorentz symmetry of the Dirac equation is the residual symmetry. Namely, this is the combined symmetry which is left after the relative breaking of some larger symmetry group.

5.6. *Lorentz Symmetry as Combined Symmetry, View from* ^{3}*He-B*

The SO_3 symmetry group for the Dirac Hamiltonian means the rotation of the coordinate frame, $r_i \rightarrow R_{ij} r_j$, where the matrix R_{ij} describes the three-dimensional rotations, which should be simultaneously accompanied by the transformation of the Dirac matrices, $\alpha_\mu \rightarrow R_{\mu\nu} \alpha_\nu$ with the same matrix R. In ^{3}He these two are different symmetry operations. The rotations of the coordinate frame belong to the group $SO_3^{(L)}$ while the rotation of the Dirac matrix corresponds according to Eq. (5.7) to the $SO_3^{(S)}$ group of the spin rotations.

Above the superfluid transition the system is invariant under both groups: $SO_3^{(S)} \times SO_3^{(L)}$. In particular the Hamiltonian for the quasiparticle in the normal liquid $H = \delta_b^a \varepsilon_{\vec{k}}$ is invariant under both transformations independently. In the superfluid state this large group is broken and the superfluid state as well as the Hamiltonian for the quasiparticles are no more invariant under separate rotations. Only the combinations of these transformations comprise the residual symmetry group H for the properties of the B-phase in general (see Sec. 2.10) and for the Bogoliubov matrix in particular. The combined symmetry of the Bogoliubov equation thus reflects the combined symmetry of the ^{3}He-B equilibrium order parameter $A_{\alpha i}^{(0)} = \Delta_B \delta_{\alpha i}$, which does not change if the orbital rotation $A_{\alpha i}^{(0)} \rightarrow R_{ij}^{(L)} A_{\alpha j}^{(0)}$ of this matrix is accompanied by the equal rotation in the spin space, $A_{\alpha i}^{(0)} \rightarrow R_{\alpha\beta}^{(S)} A_{\beta i}^{(0)}$, i.e. $\mathbf{R A}^{(0)} \mathbf{R}^{-1} = \mathbf{A}^{(0)}$.

The separate orbital rotation or separate spin rotation transform the equilibrium order parameter to another degenerate state with $A_{\alpha i}^{(0)} = \Delta_B R_{\alpha i}$, where $R_{\alpha i}$ is the degeneracy parameter for the B-phase, and this changes the Dirac or Bogoliubov Hamiltonian

$$\mathbf{H} = \beta m(\vec{k})c^2 + c\alpha^\mu R_{\mu i} k^i . \tag{5.10}$$

The separate rotations however do not influence the energy of the system as well as the quasiparticle spectrum $E^2 = m^2(\vec{k})c^4 + c^2 k^2$. This just reflects the degeneracy of the ^{3}He-B vacuum resulting from the breaking of the relative spin-to-orbit symmetry. The properties of Eq. (5.10) are the same as those of Eq. (5.8), since these two vacuum states are obtained from each other by symmetry transformation.

The situation changes when we take into account that the degeneracy parameter matrix $R_{\mu i}$ in Eq. (5.10) is the dynamical Goldstone variable and therefore can be inhomogeneous in space or in time. This Goldstone field produces an external force acting on the quasiparticles. The other fields appear when one takes into account the non-Goldstone components of the B-phase order parameter. Let us consider for example the collective variables, which correspond to the real squashing modes in Eq. (4.27), the modes with quantum numbers $J = 2$, $T = +$. They are described by the real symmetric traceless deviations u_{ik} from the equilibrium in Eq. (4.25). For the real matrix $A_{\mu i}$ the Bogoliubov Hamiltonian (5.10) is modified to become

$$\mathbf{H} = \beta m(\vec{k})c^2 + c\alpha^\mu A_{\mu i} k^i / \Delta_B , \qquad (5.10a)$$

and the square of this Hamiltonian gives the following quasiparticle energy

$$E_{\vec{k}}^2 = m(\vec{k})^2 c^4 + g^{ij} k_i k_j ,$$

where the effective metric tensor is expressed in terms of the real squashing collective variables:

$$g^{ij} = c^2(\delta^{ij} + 2u^{ij} + u^{il} u^{lj}) .$$

This means that this collective mode with $J = 2$ plays the part of the gravity field for the Dirac particles.

5.7. Breaking of the Relative Lorentz Symmetry

Inspired by the analogy one may thus introduce, instead of one Lorentz group L, two different Lorentz groups. One of them is pure orbital, $L^{(L)}$, which is responsible for the 4-dimensional rotations $x_i \rightarrow L_{ij}^{(L)} x_j$ of the space-time manifold $x_i = (\vec{r}, t) = (x_i, x_0)$. Another one, $L^{(S)}$, is the group of the transformations of Dirac matrices $\gamma_i = (i\beta\vec{\alpha}, \beta)$, which is just the 4-dimensional generalization of the spin rotation group: $\gamma_i \rightarrow L_{ij}^{(S)} \gamma_j$.

Then one may suggest that in the real vacuum of particle physics the relative symmetry is broken in the same way as in the ³He-B. This means that above some critical temperature T_c the vacuum is invariant under a large group which is the product of two Lorentz groups $L^{(S)} \times L^{(L)}$, spin and orbital. In particular the equations for the fermionic excitations of this highly symmetric vacuum should be invariant under this group, e.g. the quasiparticles can obey the Klein-Gordon equation. Below the transition the large group is spontaneously broken. And the residual symmetry of the new vacuum is just the combined Lorentz group $L^{(S+L)}$. The broken relative symmetry $L^{(S-L)}$ gives rise to the Dirac equation for excitations of this new vacuum, which is thus invariant only under the combined group. This produces an effective spin orbital coupling between orbital and spin variables via the order parameter 4×4 matrix. Some non-Goldstone components of this matrix form the dynamical metric tensor and thus play the part of the gravity fields.

5.8. *Gap Nodes in the Quasiparticle Spectrum of ³He-A*
Class of Intermediate Superfluids

Substituting the equilibrium order parameter for the A-phase, Eq. (2.10), into Eq. (5.3), one obtains the following Bogoliubov matrix for the A-phase fermions:

$$\mathbf{H} = \tau_3(\epsilon(\vec{k}) - \mu) + \frac{\Delta_A}{k_F}(\vec{\sigma} \cdot \hat{d})(\tau_1\hat{e}^{(1)} \cdot \vec{k} - \tau_2\hat{e}^{(2)} \cdot \vec{k}) , \qquad (5.11)$$

and the following quasiparticle spectrum, which again comes from the diagonal $\mathbf{H}^2$:

$$E_{\vec{k}}^2 = (\epsilon(\vec{k}) - \mu)^2 + \Delta_A^2 \frac{(\vec{k} \times \hat{l})^2}{k_F^2} \approx v_F^2(k - k_F)^2 + \Delta_A^2(\hat{k} \times \hat{l})^2 . \qquad (5.12)$$

The most prominent feature of this spectrum is that the energy of quasiparticles becomes zero at two points in the momentum space, at $\vec{k} = \pm k_F\hat{l}$ (Fig. 5.2). These two Fermi points lead to many specific features in the low temperature dynamics of the liquid, since now the massless fermions are not exponentially frozen out, but can be easily created in dynamical processes.

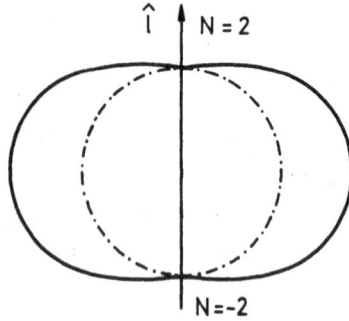

Fig. 5.2. Gap nodes in the quasiparticle spectrum in the A-phase (Fermi points). At these points, $\vec{k} = \pm k_F \hat{l}$, the quasiparticle spectrum crosses the chemical potential. Most of the exotic properties of the A-phase are related to these nodes.

In particular a massive external object, moving in the liquid with the arbitrary small velocity $\vec{v}$, always radiates the quasiparticles. The same object, moving in ^{3}He-B, radiates fermions only starting from some critical velocity $v = v_L$, known as Landau velocity, at which the emission is allowed by the energy and momentum conservation. This threshold corresponds to such a velocity $\vec{v}$ at which the energy of the excitation in the coordinate frame, moving with the massive object, $E_{\vec{k}} - \vec{k} \cdot \vec{v}$, first becomes negative. For the ^{3}He-B with the quasiparticle spectrum in Eq. (5.6), the Landau critical velocity is $v_L = \Delta_B / k_F$, while for ^{3}He-A, $v_L = 0$, i.e., the moving object always experiences friction force from the liquid.

Nevertheless the A-phase is still superfluid since it can flow without friction. The possibility to emit the quasiparticles does not prevent the superflow to circulate without friction, which is a consequence of the broken gauge invariance, rather than the property of the quasiparticle spectrum. The emitted fermionic quasiparticles fill all the negative energy levels $E_{\vec{k}} - \vec{k} \cdot \vec{v} < 0$, and after that further emission stops due to the Pauli exclusion principle for fermions. This is different from the case of a moving massive object which occupies only a finite part of the (infinite) volume of the liquid, therefore the quasiparticles emitted by the moving object cannot fill all the negative levels. As a result the emission process by a moving object is never stopped and the moving object experiences the friction force from the liquid.

So one may say that the superfluid ^{3}He-A belongs to the class of the intermediate liquids between two extreme cases. On one pole there are classical superfluid liquids like ^{3}He-B and superfluid ^{4}He, which are superfluid in both senses: they sustain the persistent flow of liquid without friction and the external body also can move without any dissipation. On the other pole is the normal viscous liquid, where no dissipationless regime takes place. In the A-phase the liquid moves without friction, while the external body motion is dissipative.

5.9. *Combined Gauge Symmetry and Gap Nodes*

The unique phenomenon of nodes in the quasiparticle gap (Fermi points) is not accidental but results completely from the combined gauge-orbit symmetry of the A-phase state, $U(1)^{\text{combined}} = U(1)SO_2^{(L)}$. To see this, let us apply to the off-diagonal element $\Delta_b^a(\vec{k})$ the combined symmetry operation which consists of the gauge transformation by the parameter α and the orbital rotation $\mathbf{R}_{(\hat{l},2\alpha)}^{(L)}$ about axis $\hat{l}$ by the angle 2α. Since this operation belongs to the residual symmetry H of the A-phase vacuum state, it leaves the order parameter invariant:

$$\Delta(\vec{k}) = e^{2i\alpha}\Delta(\mathbf{R}_{(\hat{l},2\alpha)}^{(L)}\vec{k}) \ . \tag{5.13}$$

Let us now choose the momentum $\vec{k}$ along $\hat{l}$, then the momentum is invariant under the orbital rotation $\mathbf{R}_{(\hat{l},2\alpha)}^{(L)}$, i.e. $\mathbf{R}_{(\hat{l},2\alpha)}^{(L)}\vec{k} = \vec{k}$ and one has

$$e^{2i\alpha}\Delta(\mathbf{R}_{(\hat{l},2\alpha)}^{(L)}\vec{k}) = e^{2i\alpha}\Delta(\vec{k}) \quad \text{for} \quad \vec{k} \parallel \hat{l} \ . \tag{5.14a}$$

From Eqs. (5.13), (5.14a) it follows that for $\vec{k} \parallel \hat{l}$ one has

$$\Delta(\vec{k} \parallel \hat{l}) = e^{2i\alpha}\Delta(\vec{k} \parallel \hat{l}) \tag{5.14b}$$

for arbitrary α. Therefore the off-diagonal matrix element disappears, $\Delta(\vec{k} \parallel \hat{l}) = 0$, and the energy gap is zero at $\vec{k} \parallel \hat{l}$, so $E_{\vec{k}=\pm k_F\hat{l}} = 0$.

The important point in this derivation is that the residual symmetry H in the A-phase contains the gauge symmetry (in combination with the orbital symmetry group) which gives rise to the phase factor in Eq. (5.14b). So one may conclude that the nodes may appear if the gauge symmetry is nontrivially broken, i.e. if the pure gauge symmetry is broken in such a way that the residual symmetry H of the system contains the elements of the gauge group in combination with the elements of some other group. If such breaking of symmetry occurs in superconductors, we call this unconventional superconductivity. It is possible that similar unconventional superconductivity, which gives rise to the Fermi points or Fermi lines in the superconducting state, takes place in heavy fermion compounds.

5.10. *Stability of Gap Nodes in the A-phase.*
Evolution of Fermi Points at $A \to B$ Transition

The problem now arises as to what occurs with the nodes in the gap when the symmetry which leads to their existence is externally violated. At first glance the nodes should perish since the main reason for their existence seems to have been removed. However, as we shall see later, the answer is quite unexpected, and the A-phase proves to be unique in this respect: the Fermi points in the A-phase do not disappear for any (not very large) perturbation of the order parameter, including even such perturbations at which the superfluid state is no more described by the A-phase order parameter. To illustrate this let us consider what occurs with the zeroes when the A-phase is transformed to the B-phase by continuous modification of the order parameter. This is not an academic problem, since such transformation of the order parameter takes place in real space within the AB interface, whose structure was considered above, in Sec. 3.

The positions of the gap nodes as a function of the coordinate x which intersects the AB interface is found by substitution of the solution for the order parameter $A_{\alpha i}(x)$ into Eq. (5.3) for the gap function and calculating the determinant of this 4×4 matrix. The zeroes of the determinant correspond to the gap nodes. The result of the calculations is shown in Fig. 5.3.

The evolution of nodes occurs as if there were some conservation of the charge related to the gap nodes. According to this one should ascribe a charge +2 to the node at the north pole ($\vec{k} = k_F \hat{l}$) and −2 to the node at the south pole ($\vec{k} = -k_F \hat{l}$). When x approaches the AB-wall the double nodes split into pairs of single nodes ($\vec{k} = \pm k_F \hat{l}_1$ and $\vec{k} = \pm k_F \hat{l}_2$). The single nodes move along the meridianal line in the plane $k_x = 0$ until the nodes with opposite charges annihilate each other.

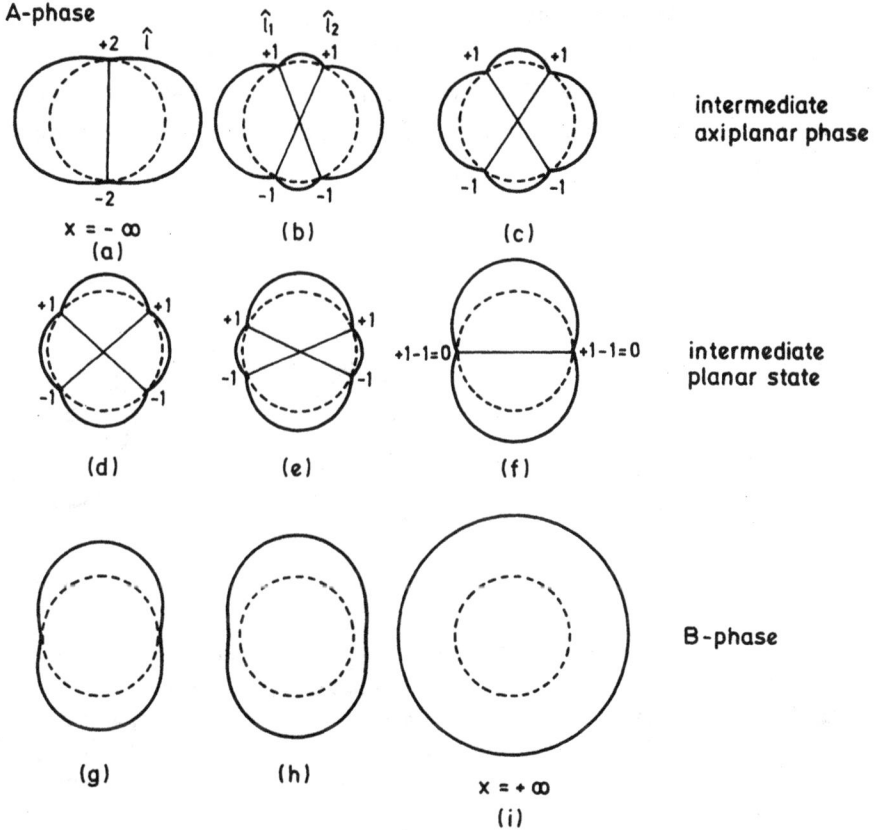

Fig. 5.3. Illustration of the high stability of the Fermi points. Evolution of the topologically stable Fermi points, the gap nodes, when the system continuously transforms from the A-phase on one side of the AB interface to the B state with isotropic gap on the other side. The stability is guaranteed by the conservation of the topological charge N discussed in Sec. 9. This charge can be eliminated only by annihilation with an opposite charge. The intermediate superfluid state, where annihilation takes place, is the planar phase, discussed later in Sec. 8.

Further in Sec. 9 we show that the corresponding charge of the Fermi points in the A-phase has a topological origin. So if this charge is nonzero, one cannot destroy the isolated node by perturbation of the order parameter: only annihilation with the node of the opposite charge is possible. Starting from the powerful topological stability of the point nodes in the A-phase, one may conjecture that the stable point nodes may persist even in the conventional superconducting state: if the pair (node-antinode) appears accidentally at some value of strong Cooper pairing potential, the nodes will exist above this threshold, see Fig. 5.4.

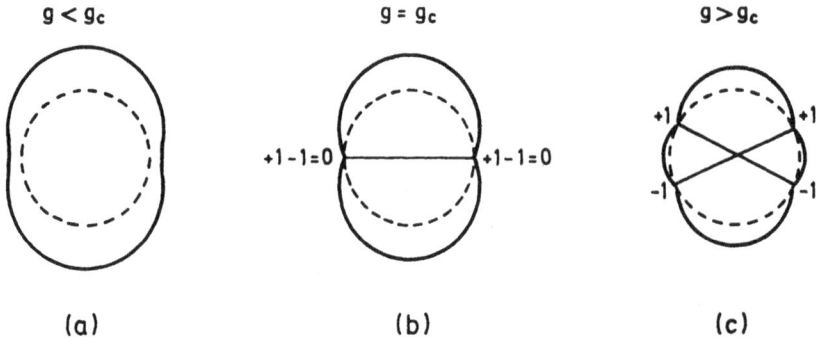

Fig. 5.4. The topologically stable nodes may persist even in conventional superconductors, if the pairing interaction parameter g is strong enough. The nodes can appear as bifurcation, or zero temperature Lifshitz transition, at some critical value of g. a) State without gap nodes, b) bifurcation point, at which pairs of Fermi points with opposite charges first appear, c) state with gap nodes.

5.11. *Spectrum Near the Fermi Points and Relativistic Massless Particles*

In the vicinity of the nodes the quasiparticle spectrum may be expanded in terms of the small parameter $\vec{k} - (\pm)k_F\hat{l}$. In the main approximation the square of the spectrum, $E_{\vec{k}}^2$ in Eq. (5.12) is the quadratic form of the deviations. It is convenient to write it in the following symmetric way, which will be useful later:

$$E_{\vec{k}}^2 = g^{ij}(k_i - eA_i)(k_j - eA_j) . \tag{5.15}$$

Here $\vec{A} = k_F\hat{l}$, $e = +1$ for excitations near the north pole, i.e. near $\vec{k} = k_F\hat{l}$, and $e = -1$ for the excitations near the south pole, i.e. near $\vec{k} = -k_F\hat{l}$, and g^{ij} is

$$g^{ij} = c_\parallel^2 \hat{l}_i \hat{l}_j + c_\perp^2 (\delta_{ij} - \hat{l}_i \hat{l}_j) \, , \qquad (5.16)$$

with $c_\parallel = v_F$ and $c_\perp = \Delta_A/k_F$.

So in the vicinity of the poles the spectrum becomes relativistic, i.e. the symmetry of the system is enhanced in the low energy limit. As a result the fermions in the A-phase near the gap nodes are equivalent to the relativistic massless charged particles with the electric charge e, moving in electromagnetic and gravitational fields. The effective electromagnetic field $\vec{A}$ has nothing to do with the $U(1)$ gauge field of the electrically charged system, this is some new dynamical field produced by the influence of the order parameter textures on the Fermi excitations. The role of the vector potential $\vec{A}$ of the electromagnetic field is played here by the Goldstone $\hat{l}$ field. If $\hat{l}$ changes in space and time, the effective magnetic field $\vec{B} = k_F \vec{\nabla} \times \hat{l}$ and electric field $\vec{E} = k_F \partial_t \hat{l}$ produce the forces on the quasiparticle, which correspond to the action of real magnetic and electric fields on electrons and positrons.

Note the essential difference between the conventional electromagnetic field in the electrically charged superfluids (superconductors) and the effective electromagnetic field, produced by the order parameter. Conventional field in superconductors acts on the Bogoliubov quasiparticles somewhat unusually, because these quasiparticles represent the mixture of the particle and hole states, which transform differently under the $U(1)$ gauge transformation. This is the result of the broken $U(1)$ gauge symmetry. As a result the electromagnetic field leads to the shift of the quasiparticle energy, $E_{\vec{k}} \to E_{\vec{k}} - \vec{k} \cdot \vec{A}$. On the contrary the effective electromagnetic field, produced by the order parameter, acts on the Bogoliubov quasiparticles as some new $U(1)$ gauge field. The field shifts the momentum, $\vec{k} \to \vec{k} - \vec{A}$, in complete analogy with gauge invariance. This new gauge invariance holds only for the fermionic fields and only in the vicinity of the Fermi points, nevertheless it appears to be important for many anomalous properties of the A-phase related to the existence of the gap nodes (see Sec. 6).

As is seen from Eqs. (5.15) and (5.16) and also from the next subsection, the gravity field is also modeled here by the degeneracy parameter which produces the local metric for the quasiparticles, described by the metric tensor g^{ij}. The other components of the degeneracy parameters as well as the non-Goldstone components of the order parameter also give rise to the fields acting on the quasiparticles. So the gauge fields in the superfluid ^{3}He result from the inhomogeneous distribution of the order parameter and from the order parameter fluctuations. This analogy allows us to simplify many complicated calculations in the A-phase, by just reducing them to the solved problems in quantum electrodynamics.

One moment which should be pointed out here is that the symmetry of the quasiparticle spectrum is enhanced both in the low-energy and high-energy limits. Symmetry enhancement in the high-energy limit is common both for condensed matter and particle physics, since in this limit the symmetry breaking, which occurs as the low energy phenomenon, is already ineffective and the system has effectively the initial unbroken symmetry G. While the symmetry enhancement in the low-energy limit seems to be the privilege of condensed matter only: the complicated form of the spectrum in solid state physics always becomes simplified in the vicinity of some symmetry points. This just occurs in the A-phase near the gap nodes, where the quasiparticle spectrum becomes linear and, as a result, relativistic.

5.12. *Left-handed and Right-handed Fermions Near the Fermi Points*

The square of the energy in Eq. (5.15) does not give complete information on the spectrum. Instead of investigating the square of the quasiparticle energy in the vicinity of nodes, let us expand directly the Bogoliubov Hamiltonian $\mathbf{H}$, Eq. (5.11). The projection of the spin $\vec{S} = \frac{1}{2}\vec{\sigma}$ of the quasiparticle on the quantization axis $\hat{d}$ is a qood quantum number, since it commutes with Hamiltonian $\mathbf{H}$ in Eq. (5.12). This is in accordance with the classification of the excitations of the ^{3}He-A vacuum with M_S as one of the quantum numbers, $M_S = \pm 1/2$ for the fermions. So let us consider the spectrum with given spin projection M_S, in this case the Bogoliubov Hamiltonian is reduced to the 2×2 matrix:

$$\mathbf{H} = \tau_3(\epsilon(\vec{k}) - \mu) + \frac{2\Delta_A}{k_F} M_S(\tau_1 \hat{e}^{(1)} \cdot \vec{k} - \tau_2 \hat{e}^{(2)} \cdot \vec{k}) . \tag{5.17a}$$

At $\vec{k} = \pm k_F \hat{l}$ all the elements of $\mathbf{H}$ are zero, therefore in the vicinity of the nodes $\mathbf{H}$ is a linear function of $\vec{k} - ek_F \hat{l}$. The general form of the linear term for the 2×2 matrix is

$$\mathbf{H} = \sum_{N=1}^{3} \tau_N e_N^j (k_j - eA_i) , \tag{5.17b}$$

where the coefficients e_N^i for the given $\mathbf{H}$ in Eq. (5.17a) are

$$e_3^i = ev_F \hat{l}_i , \quad e_1^i = 2M_S \frac{\Delta_A}{k_F} \hat{e}_i^{(1)} , \quad e_2^i = -2M_S \frac{\Delta_A}{k_F} \hat{e}_i^{(2)} . \tag{5.18}$$

These coefficients e_N^i form the so-called *dreibein*, or triad, the local coordinate frame for the fermionic particles. As distinct from the initially introduced dreibein for the degeneracy parameters which consists of the unit orthogonal vectors $\hat{e}^{(1)}$, $\hat{e}^{(2)}$ and $\hat{l}$, the vectors e_N^i are no longer unit, and as we shall see below they are not necessarily orthogonal in the general case of arbitrary change of the order parameter. They are the 3-dimensional equivalent of the *fierbein* or tetrads, which are used for the description of the gravity field in the tetrad formalism of general relativity. The conventional metric tensor, Eq. (5.16), is expressed in terms of triad e_N^i as

$$g^{ij} = \sum_{N=1}^{3} e_N^j e_N^i . \tag{5.19}$$

Equation (5.15) is obtained from Eq. (5.17b) just by taking the square of both sides of Eq. (5.17b).

In particle physics the Hamiltonian, Eq. (5.17b), for gapless (massless) particles is known as Weyl Hamiltonian for the chiral particles with spin 1/2, such as neutrinos. This means that the Bogoliubov isospin $\vec{\tau}$ in the particle-hole space of superfluid ^{3}He-A plays the part of the conventional spin $\vec{\sigma}$ in particle physics. In the isotropic vacuum of particle physics, with $e_N^i = \pm c\delta_N^i$, one has

$$\mathbf{H} = \pm c\, \vec{\tau} \cdot (\vec{k} - e\vec{A}) , \tag{5.20}$$

where c is the velocity of light and signs $+$ and $-$ correspond to the chirality of the particle. The massless particles have well-defined chirality, the projection of the spin on the direction of the particle momentum. The $+$ sign corresponds to the right-handed particles with the energy $E_{\vec{k}} = c\vec{\tau} \cdot \vec{k}$. For these particles $\vec{\tau}$ is parallel to $\vec{k}$, while the $-$ sign corresponds to the left-handed particles with the energy $E_{\vec{k}} = -c\vec{\tau} \cdot \vec{k}$, for which $\vec{\tau}$ is antiparallel to $\vec{k}$.

The same takes place for the A-phase fermions, where the chirality is the projection of the Bogoliubov isospin. The chirality C here may be defined as the sign of the determinant of the *dreibein* matrix $C = \text{sign}(\text{Det}\{e_N^i\})$. Using Eq. (5.18) for e_N^i one obtains $C = -e$, i.e. the chirality of the fermion is opposite to its *electric* charge. The positively charged fermions near the north pole are left-handed, while the negatively charged fermions near the south pole are right-handed.

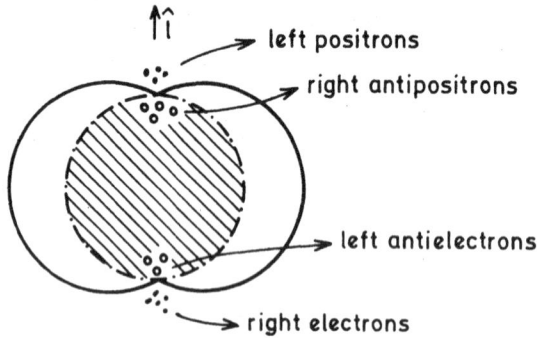

Fig. 5.5. The Bogoliubov Hamiltonian near the gap nodes is equivalent to the Weyl Hamiltonian for the electrically charged chiral particles with the chirality opposite to the sign of electric charge. So the quasiparticles in the vicinity of the north pole have positive electric charge and negative chirality (left-handed positrons), the quasiparticles in the vicinity of the south pole have negative electric charge and positive chirality (right-handed electrons). The quasiholes correspond to their antiparticles.

So many properties of the A-phase have analogy in particle physics, though they may be described in a different language. The oscillations of the $\hat{l}$ vector, orbital waves with the spectrum $\omega^2_{\text{photon}}(\vec{q}) = c^2_{\parallel}(\vec{q} \cdot \hat{l})^2$ (see Sec. 6.14, we neglect here the term with $c_{\perp}$ since the superfluid ^{3}He-A is highly anisotropic, with $c_{\perp}/c_{\parallel} \sim T_c/\epsilon(k_F) \ll 1$) correspond to the electromagnetic waves in particle physics according to the relation $\vec{A} = k_F \hat{l}$. Then, as we discussed above, the Bogoliubov isospin in ^{3}He-A corresponds to conventional spin and, as we shall see in the following subsection, vice versa, the conventional spin of the ^{3}He atom corresponds to the isospin in the weak interactions theory.

5.13. Spin-orbit Waves and W Bosons

According to the relation $\vec{A} = k_F \hat{l}$, the Goldstone field $\hat{l}(\vec{r}, t)$ results in the dynamic $U(1)$ gauge field for the fermions. Another dynamic gauge field $SU(2)$ arises if one takes into account the spin structure of the Bogoliubov-Weyl Hamiltonian and the collective spin-orbit modes of the order parameter. Again this effective gauge field has nothing to do with the rotational groups which exist in normal ^{3}He, it appears in superfluid ^{3}He-A just because the interaction of some components of the order parameter with the Bogoliubov fermions in the vicinity of the Fermi points imitates the interaction of the fermions with the gauge field. To visualize this let us fix for simplicity the orientation of the equilibrium order parameter of the A-phase in the form of Eq. (2.5) with $\hat{d}^0 = \hat{l}^0 = \hat{z}$, and consider such deviations $\delta A_{\alpha i}$ from the equilibrium state, which correspond to the orbital and spin-orbital waves in Eq. (4.24). These are the collective variables which have the orbital component z: $u_{\alpha z}$ and $v_{\alpha z}$. Substitution of these components in the Hamiltonian (5.1), (5.3) gives for $\mathbf{H}$ the following expression at momenta $\vec{k}$ in the vicinity of the nodes:

$$\mathbf{H} = ec_{\parallel}\tau_3(k_z - ek_F) + c_{\perp}\tau_1\sigma_z\big(k_x + ek_F(u_{zz} - \tau_3\sigma_x v_{yz} + \tau_3\sigma_y v_{xz})\big)$$

$$-c_{\perp}\tau_2\sigma_z\big(k_y + ek_F(v_{zz} - \tau_3\sigma_y u_{xz} + \tau_3\sigma_x u_{yz})\big) . \qquad (5.21)$$

From Eq. (5.21) it follows that these collective variables play the part of the gauge fields. The components of the orbital wave modes u_{zz} and v_{zz} comprise the deviation of the orbital vector $\delta\hat{l}$ from its equilibrium position $(u_{zz} = \delta\hat{l}_x = A_x/k_F, v_{zz} = \delta\hat{l}_y = A_y/k_F)$ and therefore form the components of the $U(1)$ gauge field $\vec{A}$. The other 4 components of the collective modes, the spin-orbit waves, comprise the components of the $SU(2)$ gauge field with the generators of the $SU(2)$ group being $\tilde{\sigma}_x = \tau_3\sigma_x$, $\tilde{\sigma}_y = \tau_3\sigma_y$ and $\tilde{\sigma}_z = \sigma_z$, since the matrices $\tilde{\sigma}_z = \sigma_z$, $\tilde{\sigma}_x = \tau_3\sigma_x$ and $\tilde{\sigma}_y = \tau_3\sigma_y$ satisfy the regular commutative rules $[\tilde{\sigma}_i, \tilde{\sigma}_j] = 2ie_{ijk}\tilde{\sigma}_k$. Further we refer to this $SU(2)$ group as $SO_3^{(\tilde{S})}$ to distinguish it from the conventional rotational groups $SO_3^{(S)}$ and $SO_3^{(L)}$.

As a result Eq. (5.21) may be rewritten in the form with explicit dependence on the $U(1)$ gauge field $\vec{A}$ and $SU(2)$ gauge field $\vec{W}^\alpha$:

$$\mathbf{H} = \sum_N e_N^i \tau_N (k_i - eA_i - e\tilde{\sigma}_\alpha W_i^\alpha) , \qquad (5.22)$$

with the following relations between W_i^α and $\delta A_{\alpha i}$:

$$W_x^\alpha + iW_y^\alpha = i(k_F/\Delta_A)e^{\alpha\beta\gamma}\hat{d}^\beta \delta A_{\gamma i} . \qquad (5.23a)$$

So only 4 components of W_i^α are nonzero:

$$W_x^x = v_{yz}k_F , \quad W_x^y = -v_{xz}k_F , \quad W_y^x = -u_{yz}k_F , \quad W_y^y = u_{xz}k_F . \quad (5.23b)$$

Equation (5.22) is equivalent to the Hamiltonian for the fermions in the standard Weinberg-Salam theory of electroweak interactions, where the $SU(2)$ group is the group of isospin rotations. So the isospin in particle physics corresponds to the slightly modified nuclear spin $\tilde{\sigma}$ of the ^{3}He atoms. The fields $\vec{W}^\alpha$ thus represent the weak bosons. The only nonzero components of the fields $\vec{W}^\alpha$ are those which are perpendicular to $\hat{d}$, i.e. with $\vec{W}^\alpha d^\alpha = 0$. This means that they correspond to the W intermediate bosons. Also from this observation it follows that the magnetic anisotropy axis $\hat{d}$ corresponds to the axis of the quantization of the weak isospin in particle physics, i.e. this is the direction of the Higgs field in the Weinberg-Salam model. So one has another analogy between the A-phase and particle physics: the $\hat{d}$ field vs. Higgs field.

5.14. *Mass of the W Bosons is Zero in the BCS Theory of* ^3He-A

In the standard theory of electroweak interactions the $SU(2)$ gauge field oscillations (the W bosons) initially have no mass and acquire the mass only after symmetry breaking, due to the Higgs mechanism. The situation in superfluid ^{3}He-A is not the same. Different from particle physics, liquid ^{3}He, initially above the transition temperature T_c, has no local $SU(2)$ (or $SO_3^{(S)}$) gauge invariance, and therefore no W bosons. The gauge fields for the fermions appear only below T_c in the ^{3}He-A state, arising as some collective motion of the order parameter: the $SU(2)$ gauge fields are the components of the spin-orbit waves. The latter are not Goldstone fields for ^{3}He-A, therefore the gauge bosons, W bosons, when they appear below T_c immediately acquire the finite mass. Nevertheless incidentally the problem of the W boson mass in the ^{3}He-A proves to be quite intriguing. If one applies the traditional Bardeen-Cooper-Schrieffer (BCS) approach to the description of Cooper pairing and uses the Bogoliubov matrix (5.11) for the fermions in equilibrium A-phase, one obtains zero mass for the W boson. Moreover the spectrum of the W bosons exactly coincides with that of photons (orbital-waves), $\omega_W^2(\vec{q}) = c_\parallel^2 (\vec{q} \cdot \hat{l})^2$, within this model Hamiltonian.

5.15. *Hidden Symmetry in the A-phase*

This paradox results from the specific hidden symmetry of the Bogoliubov Hamiltonian (5.11), which we discuss now. Usually an additional (hidden, or approximate) symmetry of a system comes from the existence of some small parameter. When this parameter is treated as zero, the hidden symmetry is exact. For example in ^{3}He the smallness of the spin-orbit coupling gives rise to a separate symmetry $SO_3^{(S)}$ of spin rotations. The small parameter for the hidden symmetry will be discussed in the next subsection.

In the superfluid ^{3}He-A the symmetry of the Bogoliubov Hamiltonian (5.11) should in principle coincide with the residual symmetry H of the A-phase vacuum, therefore it should contain two continuous Abelian groups $H = U(1) \times U(1)$, where one of the groups is the $SO_2^{(S)}$ group of spin rotations about axis $\hat{d}$ with the generator $\mathbf{S}_z$, and another one is the combined group $U(1)^{\text{combined}} = U(1)SO_2^{(L)}$ with the generator $\mathbf{Q} = \frac{1}{2}\mathbf{I} - \mathbf{L}_z$ in Eq. (4.20). It can be checked that these groups really leave the Hamiltonian (5.11) invariant, they correspond to the following two transformations of the Hamiltonian respectively:

$$e^{i\frac{1}{2}\sigma_z\theta_z}\mathbf{H}e^{-i\frac{1}{2}\sigma_z\theta_z} = \mathbf{H} , \qquad (5.24a)$$

$$e^{i\frac{1}{2}\tau_3\alpha}\mathbf{H}(\mathbf{R}^{(L)}_{(\hat{z},\alpha)}\vec{k})e^{-i\frac{1}{2}\tau_3\alpha} = \mathbf{H}(\vec{k}) . \qquad (5.24b)$$

However in addition there are two more continuous transformations which also do not change $\mathbf{H}$, these are just rotations, induced by the generators $\tilde{\sigma}_x = \tau_3\sigma_x$ and $\tilde{\sigma}_y = \tau_3\sigma_y$, discussed in Sec. 5.13:

$$e^{i\frac{1}{2}\tilde{\sigma}_x\theta_z}\mathbf{H}e^{-i\frac{1}{2}\tilde{\sigma}_x\theta_z} = \mathbf{H} , \qquad (5.24c)$$

$$e^{i\frac{1}{2}\tilde{\sigma}_y\theta_y}\mathbf{H}e^{-i\frac{1}{2}\tilde{\sigma}_y\theta_y} = \mathbf{H} . \qquad (5.24d)$$

$|Q|$	$\tilde{S}$	D	$\tilde{P}_2$	Variables	Modes
0	1	3	$-$	$u_{zy} - v_{zx}$	sound
			$-$	$u_{yx} + v_{yy}$	spin waves
			$-$	$u_{xx} + v_{xy}$	(oscillations of $\hat{d}$)
1	1	6	$+$	v_{zz}	orbital wave (photon)
			$+$	u_{yz}	spin-orbital waves
			$+$	u_{xz}	(W bosons)
			$-$	u_{zz}	orbital wave (photon)
			$-$	v_{yz}	spin-orbital waves
			$-$	v_{xz}	(W bosons)

$$(5.25)$$

$|Q|$	$\tilde{S}$	D	$\tilde{P}_2$	Variables	Modes
0	1	3	$+$	$u_{zx} + v_{zy}$	pair-breaking modes
			$+$	$u_{xy} - v_{xx}$	
			$+$	$u_{yy} - v_{yx}$	
2	1	6	$+$	$u_{zx} - v_{zy}$	clapping modes
			$+$	$u_{yy} + v_{yx}$	
			$+$	$u_{xy} + v_{xx}$	
			$-$	$u_{zy} + v_{zx}$	
			$-$	$u_{yx} - v_{yy}$	
			$-$	$u_{xx} - v_{xy}$	

These two, together with Eq. (5.24a), form the group of three-dimensional rotations $SO_3^{(\tilde{S})}$. So the vacuum symmetry H of the A-phase proves to be extended to $H^{\text{hidden}} = SO_3^{(\tilde{S})} \times U(1)^{\text{combined}}$, and one may check that this group is no more the subgroup of the large group G. This is a typical manifestation of the hidden symmetry.

Due to the hidden symmetry the collective modes should now be classified in terms of the irreducible representations of the extended (hidden symmetry) group H^{hidden}, so they are characterized by the quantum numbers $\tilde{S}$ of the $SO_3^{(\tilde{S})}$ group and Q of the Abelian group $U(1)^{\text{combined}}$. As a result the 18 principal collective modes in the A-phase are distributed in 2 three-dimensional and 2 six-dimensional representations, all with $\tilde{S} = 1$: Here $\tilde{P}_2$ is the P_2 symmetry combined with the spin rotations by π about axis $\hat{d}$.

It follows from this table that the W bosons (spin-orbital waves) belong to the same extended multiplet of the hidden symmetry group H^{hidden} as the photons (Goldstone orbital waves). This is the representation with $\tilde{S} = 1$ and $|Q| = 1$ whose dimension is $D = 2 \times (2\tilde{S} + 1) = 6$. Therefore the W bosons have the same spectrum as photons and are also massless Goldstone bosons. The Z bosons in this classification scheme (fluctuations of the conventional singlet isotropic amplitudes of the Cooper pairing) are doubly degenerate massive bosons with quantum numbers $|Q| = 1$, $\tilde{S} = 0$ and $\tilde{P}_2 = -$.

5.16. *Origin of the W Boson Mass in ^{3}He-A*

Let us find out what is the small parameter which is neglected to result in the extension of the vacuum symmetry of the A-phase state. The symmetry H^{hidden} is exact for the Hamiltonian (5.11), however this Hamiltonian itself is some approximation: we neglected the influence of Cooper pairing on the diagonal elements of the Bogoliubov matrix. The corrections to these elements are relatively small in the parameter $\Delta_B/\epsilon(k_F) \sim T_c/\epsilon(k_F) \ll 1$, and this is precisely the small parameter which is responsible for the hidden symmetry of the A-phase state.

The corrections to the BCS theory, i.e. to the diagonal elements of the Bogoliubov matrix, externally violate the group $H^{\text{hidden}} = SO_3^{(S)} \times U(1)^{\text{combined}}$ and reduce it to the subgroup $H = SO_2^{(S)} \times U(1)^{\text{combined}}$, which is the conventional symmetry group of the A-phase. This symmetry violation restores the W boson mass:

$$\omega_W^2(\vec{q}) = m_W^2 + c_\parallel^2 (\vec{q} \cdot \hat{l})^2 . \tag{5.26}$$

This mass is however anomalously small as compared with the characteristic mass of the other non-Goldstone collective modes, such as graviton (see next subsection), since it contains an additional small factor $T_c/\epsilon(k_F) \ll 1$, i.e. $m_W/m_{\text{graviton}} \sim T_c/\epsilon(k_F)$.

So in many details there is a similarity between the low-energy behavior of the ³He-A fermions and the chiral fermions in the Weinberg-Salam model. The physical background is however quite different: in particular, as distinct from the standard model of the electroweak interactions the mass of the W bosons results not from the spontaneous breaking of the initial exact symmetry, but from the next order corrections to the BCS Hamiltonian violating the approximate hidden symmetry of the A-phase vacuum.

5.17. Gravitons in ³He-A

According to Eqs. (5.17b–19) the order parameter produces also the local coordinate frame for the moving quasiparticles, which in the case of textures depend on the space and coordinate, thus simulating the gravity field. The oscillations of this field play the part of gravitons. The closest analogy with the gravitons in the Einstein theory arises when one considers the modes propagating along the anisotropy axis $\hat{l}$, since in this case the anisotropy of the A-phase is less important.

The graviton in particle physics has spirality, the spin projection on the propagation direction, equal to ± 2, while its isospin of weak interactions is zero. In the analogy between particle physics and ³He-A, the weak isospin corresponds to the nuclear spin of the ³He atom, while the spin corresponds to the Bogoliubov spin of quasiparticles. So the collective mode which corresponds to the graviton, should have quantum numbers $Q = \pm 2$ and $M_S = 0$. According to the table in Eq. (4.24) these are clapping modes.

Substituting the deviations of the order parameter $u_{zy}+v_{zx}$ and $u_{zx}-v_{zy}$, which oscillate in the clapping modes, into Eq. (5.3) for the gap function, one obtains that these are just the oscillations of the *dreibein* about its equilibrium values given by Eq. (5.18). In the first mode the angle between the vectors e_1^i and e_2^i oscillates around its equilibrium value $\pi/2$ (see Fig. 5.6), in the second mode the ratio of the lengths of these vectors oscillates around its equilibrium value, which is 1. According to Eq. (5.19), the clapping modes are thus oscillations of g^{12} and $g^{11} - g^{22}$, which in the general relativity theory correspond to the gravitons propagating along the z axis, chosen here to be parallel to $\hat{l}$.

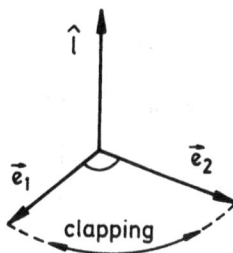

Fig. 5.6. Oscillations of the *dreibein* in one of the clapping modes, which corresponds to the graviton with the propagating g^{12} component of the metric tensor.

As distinct from the Einstein gravity, the gravitons in ^{3}He-A are not Goldstone bosons and therefore have a finite mass of the order of the gap amplitude Δ_A in the fermionic spectrum.

5.18. Cosmological Term in the Einstein Equations, View from ^{3}He-A

In the Einstein theory of gravity the mass of the graviton appears if there is a cosmological term in the Lagrangian, which has the following general form, compatible with the general covariance:

$$L_{\text{cosm}} = \Lambda_{\text{cosm}}\sqrt{-g} \ . \tag{5.27}$$

There are still some controversies related to the cosmological term. On the one hand the cosmological term, if it does not exist originally, should appear due to vacuum fluctuations. On the other hand, the simplest calculations of the vacuum fluctuations effect, i.e. response of the fermionic vacuum on the change of the metric, gives a Λ_{cosm} value many orders of magnitude (120) higher than the experimental upper limit. So it is interesting to find out what is the situation with the corresponding cosmological term in superfluid ³He.

The energy term, which explicitly depends on the components of the metric tensor, is the mass term for the clapping modes. Substituting the deviations of the order parameter $u_{zx} - v_{zy}$ and $u_{zy} + v_{zx}$ into, say, Ginzburg-Landau functional (2.1), one obtains the following equivalent of the cosmological term in ³He-A:

$$L_{\text{cosm}}^{\text{A-phase}} \sim m_{\text{graviton}}^2 [(u_{zx} - v_{zy})^2 + (u_{zy} + v_{zx})^2]$$

$$\sim m_{\text{graviton}}^2 [(g^{11} - g^{22})^2 + 4(g^{12})^2] \, . \tag{5.28}$$

The main difference with Eq. (5.27) is that this term comprises the square of the deviations of the metric tensor from its equilibrium value (5.16) in the vacuum state of the A-phase. This means that there are two metric tensors which are important: the total dynamical metrics g^{ij}, which corresponds to the gravity field, and its equilibrium value g_0^{ij} describing the reference frame of the equilibrium vacuum. The latter is given by Eqs. (5.19) and (5.16) and is defined by the orientation of the anisotropy vector $\hat{l}$. So g_0^{ij} represents the orientational degrees of freedom of the metric tensor, which depend on the space-time coordinates in the presence of the $\hat{l}$ textures. The other degrees of freedom, corresponding to the deformation of the *dreibein*, are described by the deviations δg^{ij} of the general metric tensor g^{ij} from g_0^{ij}. So gravity in the A-phase represents one of the versions of the bi-metric theory of gravitation.

The general form for the cosmological term near the equilibrium metric in the bi-metric formalism is

$$S_{\text{cosm}} \sim \delta g^{ik} \delta g^{mn} [(\tilde{m}_{\text{graviton}}^2 - m_{\text{graviton}}^2) g_{ik}^0 g_{mn}^0 + m_{\text{graviton}}^2 (g_{im}^0 g_{kn}^0 + g_{in}^0 g_{mk}^0)] \, , \tag{5.29}$$

where $\tilde{m}_{\text{graviton}}$ is the mass of an additional graviton with spirality $Q = 0$, in which the component $g^{11} + g^{22}$ of the metric tensor is oscillating. This additional graviton exists also in the A-phase, this is the pair-breaking mode $u_{xx}+v_{zy}$ in Eq.(4.24) with $Q = 0$. The ratio of the masses $\tilde{m}_{\text{graviton}}/m_{\text{graviton}}$ depends on the β parameters in the Ginzburg-Landau functional (2.1), in the BCS model it equals $\sqrt{2}$.

Different forms of the cosmological term in the Lagrangian give rise to different forms of the Einstein equation $\delta L/\delta g^{ij} = 0$. In the canonical gravity theory the cosmological term (5.27) produces the term $\Lambda_{\text{cosm}}g_{ij}$ in the Einstein equations. This term describes the gravitating mass of the vacuum, which is in the heart of the contradiction between the calculated large value of Λ_{cosm} from the quantum fluctuations and the experimental extremely small value of the vacuum mass.

In the A-phase (as well as in the B-phase) the cosmological term in the Einstein equation $\delta L/\delta g^{ij} = 0$, which follows from Eq. (5.29), is linear in $\delta g^{ij} = g^{ij} - g_0^{ij}$, and therefore disappears in the equilibrium vacuum state. This means that the vacuum itself has no gravitating mass, gravity field can be produced by the vacuum, only if there are some deviations of the vacuum from its equilibrium, e.g. textures. So within the bi-metric theory of gravity the large value of the calculated cosmological constant Λ_{cosm} is not necessary to be in disagreement with experimental cosmology.

6
Orbital Dynamics and Anomalies in Quantum Field Theory

6.1. *Lie Algebra of Poisson Brackets for A-phase Orbital Dynamics*

The orbital dynamics of the A-phase is the dynamics of the orbital part of the degeneracy parameter, the complex vector $\vec{\Psi} = \hat{e}^{(1)} + i\hat{e}^{(2)}$. The relevant group G related to the motion of this orbital *dreibein* is the group of transformations of this variable: the gauge group $U(1)$, which changes the phase of the complex vector, and the group of the orbital rotations $SO_3^{(L)}$. The corresponding generators in the classical limit are the mass density ρ and the density of the orbital momentum $\vec{L}$ of the Cooper pair rotation. If one considers the case of zero temperature, and therefore neglect the normal motion variable $\vec{v}_n$, then the corresponding Lie algebra of the Poisson brackets for the orbital (superfluid and liquid-crystal) dynamics is:

$$\left\{ L_i(\vec{r}\,'), L_j(\vec{r}\,') \right\} = -e_{ijk} L_k(\vec{r}) \delta(\vec{r} - \vec{r}\,') \ , \tag{6.1a}$$

$$\left\{ L_i(\vec{r}), \Psi_j(\vec{r}\,') \right\} = -e_{ijk} \Psi_k(\vec{r}\,') \delta(\vec{r} - \vec{r}\,') \ , \tag{6.1b}$$

$$\left\{ \rho(\vec{r}), \vec{\Psi}(\vec{r}\,') \right\} = -\frac{2m_3}{\hbar} i \vec{\Psi}(\vec{r}\,') \delta(\vec{r} - \vec{r}\,') \ . \tag{6.1c}$$

In terms of the observable physical variables, orbital anisotropy vector $\hat{l} = \frac{1}{2i} \vec{\Psi}^* \times \vec{\Psi}$ and superfluid velocity $\vec{v}_s = \frac{1}{2i} \Psi_i^* \vec{\nabla} \Psi_i$ in Eq. (3.7), which enter the London hydrodynamical energy Eq. (3.9a), the nonzero Poisson brackets are obtained from Eqs. (6.1):

$$\left\{ L_i(\vec{r}), L_j(\vec{r}') \right\} = -e_{ijk} L_k(\vec{r}) \delta(\vec{r} - \vec{r}') , \qquad (6.2a)$$

$$\left\{ L_i(\vec{r}), \hat{l}_j(\vec{r}') \right\} = -e_{ijk} \hat{l}_k(\vec{r}') \delta(\vec{r} - \vec{r}') , \qquad (6.2b)$$

$$\left\{ L_i(\vec{r}), v_{sj}(\vec{r}') \right\} = -\frac{\hbar}{2m_3} \hat{l}_i(\vec{r}') \nabla'_j \delta(\vec{r} - \vec{r}') , \qquad (6.2c)$$

$$\left\{ \rho(\vec{r}), \vec{v}_s(\vec{r}') \right\} = \vec{\nabla}' \delta(\vec{r} - \vec{r}') . \qquad (6.2d)$$

The combined gauge symmetry of the A-phase plays the most important part in the A-phase dynamics. In addition to the nonpotential superflow, it is also manifested in the Poisson brackets scheme. The result of this combined symmetry is the existence of the dynamical invariant, which we denote as C_0:

$$C_0(\vec{r}) = \rho - \frac{2m_3}{\hbar} \vec{L}(\vec{r}) \cdot \hat{l}(\vec{r}) . \qquad (6.3)$$

This is just the classical limit of the generator **Q** of the combined symmetry: $C_0 = \frac{2m_3}{\hbar} Q$, and since the combined symmetry is the residual symmetry of the vacuum state, its generator should commute with all other variables: $\left\{ C_0 , \hat{l} \right\} = \left\{ C_0 , \vec{v}_s \right\} = \left\{ C_0 , \rho \right\} = 0$. From Eqs. (6.2) it follows that it is really so and therefore C_0 is constant in time: $\partial_t C_0 = \left\{ C_0 , H \right\} = 0$.

In the simplest model of a Bose liquid consisting of the well separated molecules with the same symmetry as the Cooper pair in the A-phase, the existence of such dynamical invariant is trivial. In this model the density $\vec{L}(\vec{r})$ of the Cooper pair orbital momentum is just the orbital momentum per atom, $\frac{1}{2}\hbar\hat{l}$, times the particle density ρ/m_3, i.e., $\vec{L} = (\hbar/2m_3)\rho\hat{l}$ and therefore the dynamical invariant C_0 is identically zero. The Poisson bracket approach shows that the constraint $C_0 = 0$ is too rigid, the scheme admits a less rigid constraint: $\partial_t C_0 = 0$.

6.2. *Anomaly Cancellation as Lifshitz Transition*

Since all the anomalies come from the gap nodes (Fermi points) in the fermionic spectrum, we first derive the orbital motion equations in the anomaly-free case, where the gap nodes are absent. This may occur, say, at large pairing potential, when the Cooper pairing essentially influences the diagonal elements of the Bogoliubov matrix in such a way that the chemical potential μ in Eq. (5.11) decreases and finally, at some critical value of the pairing potential, crosses zero. Just at this moment the Fermi points annihilate each other. After this transition, when μ becomes negative, the quasiparticle energy (5.12) is nowhere zero, and all anomalies disappear. In unconventional superconductors there may be annihilation of the Fermi points from different energy bands.

Here we came across a very wide class of phase transitions, which take place only at zero temperature. The transition occurs when some other external parameter like pressure crosses the critical value at which the physical properties of the system become essentially different. A typical example is the metal-insulator transition. There is no symmetry change at the transition, but reconstruction of the properties of the system in the momentum space, such as the Fermi surface topology, occurs. In our case this is the annihilation of the Fermi points. The zero temperature phase transition with a change of the Fermi surface structure is usually called the Lifshitz transition. Later in Sec. 9 we shall discuss the Lifshitz transition with change of some internal momentum space topological invariant in relation to the quantum Hall effect and fractional statistics.

As in conventional second order phase transitions, at which some hydrodynamical parameter — the rigidity (stiffness) — appears, the Lifshitz transition may be also accompanied by the appearance or disappearance of some physical parameter. In the transition discussed here this is just the parameter C_0. C_0 is zero in the anomaly-free case, and becomes nonzero when the Fermi points first appear. We first consider the orbital dynamics in the anomaly-free case, i.e. with $C_0 = 0$. The simplest model of a Bose condensate of the well separated molecules, each with the orbital momentum $\hbar \hat{l}$, is representative of the anomaly-free state.

6.3. *The Anomaly-Free Equations for Orbital Dynamics*

The Hamiltonian for orbital dynamics with dynamical constraint $\vec{L} = (\hbar/2m_3)\rho\hat{l}$ is the zero temperature London energy of the system plus a term with the Lagrange multiplier to satisfy the constraint:

$$H = E(\rho, \hat{l}, \vec{v}_s) - \vec{\Omega} \cdot (\vec{L} - (\hbar/2m_3)\rho\hat{l}) . \tag{6.4}$$

After elimination of the Lagrange multiplier $\vec{\Omega}$ from the Liouville equations one obtains the following closed system of equations:

$$\partial_t \rho + \vec{\nabla} \cdot \vec{j} = 0 , \tag{6.5}$$

$$\partial_t \vec{v}_s = -\vec{\nabla}\mu + \frac{\hbar}{2m_3} e_{ijk} \hat{l}_i \partial_t \hat{l}_j \vec{\nabla} \hat{l}_k , \tag{6.6}$$

$$\rho \partial_t \hat{l} = \frac{2m_3}{\hbar} \frac{\delta E}{\delta \hat{l}} \times \hat{l} - (\vec{j} \cdot \vec{\nabla})\hat{l} . \tag{6.7}$$

Here $\vec{j} = \frac{\delta E}{\delta \vec{v}_s}$ is the mass current, which contains: a) the supercurrent with the superfluid density corresponding to the total mass density at zero temperature, and b) the orbital current $\frac{1}{2}\vec{\nabla} \times \vec{L}$ related to the orbital momentum:

$$\vec{j} = \rho\vec{v}_s + \frac{1}{2}\vec{\nabla} \times \left(\frac{\hbar}{2m_3}\rho\hat{l}\right) . \tag{6.8}$$

The chemical potential here is $\mu = \frac{\delta E}{\delta \rho}$.

In the stationary case Eqs. (6.5) and (6.7) transforms into Eqs. (3.20) under the equilibrium conditions of the orbital textures. Equation (6.6) ensures that the Mermin-Ho relation between $\vec{v}_s$ and $\hat{l}$ is satisfied. If one applies the *curl* operation to both sides of Eq. (6.6) one obtains the conservation of this relation in time:

$$\partial_t \left(\vec{\nabla} \times \vec{v}_s - \frac{\hbar}{2M} e_{ikl} l_i \vec{\nabla} l_k \times \vec{\nabla} l_l\right) = 0 .$$

The system of equations (6.5-7) is self-consistent, which means that the conservation laws for the energy E and for the linear momentum $\vec{j}$ are satisfied: $\partial_t E = -\vec{\nabla}_k \cdot \vec{Q}$ and $\partial_t j_i = -\nabla_k \Pi_{ik}$, with $\vec{Q}$ and Π_{ik} being the energy and momentum flux. So the low energy orbital dynamics at $T = 0$ is completely defined by the soft Goldstone variables $\hat{l}$, $\vec{v}_s$, and by the soft hydrodynamical variable ρ. No contribution comes from the incoherent degrees of freedom of the liquid, since the fermionic degrees of the anomaly free superfluid are frozen out at low temperatures. In other words, only the coherent motion of vacuum without any normal excitation is possible in the low energy dynamics. In quantum field theory this corresponds to the dynamics of the classical electromagnetic and other gauge fields when the motion is too slow to create fermionic particles from the vacuum.

Quite a different situation should take place on the other side of the Lifshitz transition where the nodes in the qusiparticle spectrum appear. In this case the incoherent motion of the created gapless fermions will be added to the system. However, even within this simplified system without gap nodes the superfluid dynamics has peculiar properties, again related to the combined gauge symmetry.

6.4. *Phase Slippage Through the Dynamics of the Orbital Vector*

Let us compare Eq. (6.6) for the superfluid velocity with the corresponding Eq. (4.35) for ³He-B and superfluid ⁴He. Equation (6.6) contains an additional driving force on the supercurrent, which is produced by the dynamics of the $\hat{l}$ texture. In this term the combined gauge symmetry of the A-phase manifests itself by intrinsical coupling of the superfluid and liquid-crystal properties of the liquid: the relaxation of the superfluid velocity can occur continuously by space and time evolution of the liquid-crystal field $\hat{l}(\vec{r}, t)$. Due to this interrelation the gauge rotation is indistinguishable from the orbital rotation θ_3 about axis $\hat{l}$, so the superfluid velocity is $\vec{v}_s = (\hbar/2m_3)\vec{\nabla}\theta_3$. The orbital rotation about axis $\hat{l}$ belongs to the non-Abelian group $SO_3^{(L)}$, and one of the important properties of this group is that any winding of the θ_3 angle may be continuously unwound by the rotations θ_1 and θ_2 about the other axes $\hat{e}^{(1)}$ and $\hat{e}^{(2)}$ which are perpendicular to $\hat{l}$, and therefore the rotation about them produces the rotation of $\hat{l}$ (see Fig. 2.1).

This is the essence of the mechanism of the supercurrent relaxation due to the $\hat{l}$ vector dynamics. Such mechanism takes place unless the $\hat{l}$ vector is pinned by the wall of the vessel. So in the thin channels, where $\hat{l}$ is heavily pinned, the continuous unwinding is impossible and the supercurrent behaves in the traditional manner, as in ³He-B, superfluid ⁴He and conventional superconductors, where unwinding occurs through the motion of the singular vorticity or at the phase-slip centers while in the broad channels this new mechanism of continuous textural dynamics dominates. This dynamics becomes a steady state if the textural relaxation of the supercurrent is compensated by the external driving force $\vec{\nabla}\mu$,

$$\vec{\nabla}\mu = \langle \frac{\hbar}{2m_3} e_{ijk}\hat{l}_i\partial_t\hat{l}_j\vec{\nabla}\hat{l}_k \rangle \ ,$$

so the velocity level is maintained, while the $\hat{l}$ vector performs the Josephson oscillations. Such periodic motion of the texture under external temperature gradient was observed in the ultrasonic experiments, due to the high sensitivity of the sound absorption to the $\hat{l}$ vector orientation with respect to the direction of sound propagation.

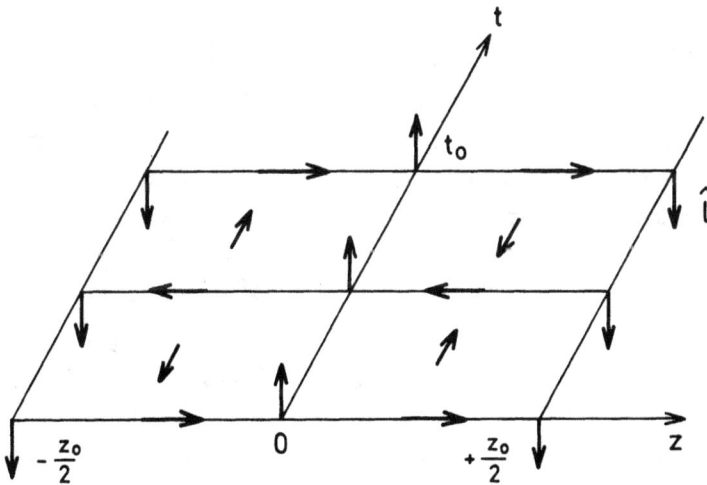

Fig. 6.1. Instanton-pseudoparticle in the 2-dimensional space-time plane, which describes the process of transition from one vacuum state to another. The vacuum states at $t = 0$ and $t = t_0$ have the same distribution of the $\hat{l}$ field, but differ by the winding number of the θ_3 angle: $m_{\Phi}(t = t_0) - m_{\Phi}(t = 0) = 2$. So this process results in the unwinding of m_{Φ}. The lattice of instantons – periodic in space and time oscillations of the $\hat{l}(z,t)$ texture – represents the ac Josephson effect in ^{3}He-A.

As an illustration of this mechanism let us consider the following $\hat{l}(z,t)$ texture, periodic in space and time with the periods z_0 and t_0 respectively, which gives rise to a constant relaxation rate of the superfluid velocity in the z direction (see Fig. 6.1):

$$\hat{l}(z,t) = \hat{z}\cos\beta(z) + \sin\beta(z)\big(\hat{x}\cos\alpha(t) + \hat{y}\sin\alpha(t)\big) \ . \qquad (6.9)$$

Within the period, say at $-z_0/2 < z < z_0/2$ we choose

$$\cos\beta(z) = 1 - 4\frac{|z|}{z_0}\ , \quad \alpha(t) = 2\pi\frac{t}{t_0}\mathrm{sign}\ z\ , \qquad (6.10)$$

Then according to Eq. (6.6) the textural relaxation of the velocity is expressed in terms of the time and space periods of the texture:

$$\partial_t(\vec{v}_s)_z + \partial_z\mu = \frac{2h}{m_3 t_0 z_0}\ . \qquad (6.11)$$

The kinetic energy of the superflow is converted into heat due to the frictional resistance to the motion of the orbital vector $\hat{l}$ from the normal component of the liquid. In the steady state flow, with $\langle\vec{v}_s\rangle = $ const, the periods z_0 and t_0 of the texture are defined by the external driving force and by the friction force parameter.

6.5. Gap Nodes Contributions to the Orbital Dynamics

Other new phenomena appear when the Lifshitz transition takes place. At this transition the chemical potential μ becomes positive and two Fermi points appear at $\vec{k} = \pm\tilde{k}\hat{l}$, where $\tilde{k}^2 = 2m_3\mu$. Microscopic calculations show that at this moment the orbital equations change drastically. To the equations for the current $\vec{j}$ and orbital momentum $\hat{l}$, additional terms are added which may be described in terms of the unique parameter

$$C_0 = \frac{\tilde{k}^3}{3\pi^2}\ , \qquad (6.12)$$

which proves to be the dynamical invariant in Eq. (6.3).

This parameter thus modifies the effective dynamical value of the orbital angular momentum of the Cooper pair, which is now reduced to the value $\vec{L} = (\hbar/2m_3)(\rho - m_3 C_0)\hat{l}$. Near the Lifshitz transition this reduction is small, but far from the transition in the real A-phase where $\tilde{k} \approx k_F$, the C_0 parameter practically completely eliminates the magnitude of the Cooper pair dynamical angular momentum, since $\rho \approx m_3 k_F^3/3\pi^2 \approx m_3 C_0$.

The dynamical equations are obtained using now the modified constraint $\vec{L} = (\hbar/2m_3)(\rho - m_3 C_0)\hat{l}$, which results in the following modification of Eq. (6.7) for $\hat{l}$:

$$(\rho - m_3 C_0)\partial_t \hat{l} = \frac{2m_3}{\hbar} \frac{\delta E}{\delta \hat{l}} \times \hat{l} - (\vec{j} \cdot \vec{\nabla})\hat{l} \ . \tag{6.13}$$

In addition the calculated London energy $E(\rho, \hat{l}, \vec{v}_s)$ is now different, which gives the following correction to the mass current

$$\vec{j} = \frac{\delta E}{\delta \vec{v}_s} = \rho \vec{v}_s + \frac{1}{2}\vec{\nabla} \times (\frac{\hbar}{2m_3}\rho\hat{l}) - \frac{\hbar}{2}C_0\hat{l}(\hat{l} \cdot \vec{\nabla} \times \hat{l}) \ . \tag{6.14}$$

The last term in Eq. (6.14) is the so-called anomalous current discussed in Sec. 6.10.

The most surprising consequence of all these modifications is that the linear momentum of the vacuum motion is no more conserved. If using Eqs. (6.5), (6.6), (6.13) and (6.14) one calculates the time derivative of the mass current (which is simultaneously the linear momentum of liquid), one obtains:

$$\partial_t j_i + \nabla_k \Pi_{ik} = -\frac{3}{2}C_0\hat{l}_i(\partial_t \hat{l} \cdot \vec{\nabla} \times \hat{l}) \ . \tag{6.15}$$

The nonconservation of the linear momentum of the Bose-condensate and the reduction of the effective dynamical angular momentum of Cooper pairs of the Bose-condensate result from the opening of the nodes (Fermi points) in the quasiparticle spectrum. These are the windows where the quasiparticles are created taking away the linear and angular momenta from the coherent vacuum motion.

6.6. Anomaly in Orbital Dynamics and Chiral Anomaly

As we have seen in Sec. 5, in the vicinity of the Fermi points the quasiparticles, interacting with the vacuum motion, may be described in terms of the chiral fermions interacting with the gauge fields. Therefore the anomaly in the orbital dynamics may be described in terms of the chiral anomaly, extensively investigated in particle physics. The chiral anomaly in the physics of massless chiral fermions, obeying the Weyl equation (5.20), $\mathbf{H} = C\, c\vec{\tau}\cdot\vec{k}$ (where $C = \pm$ is the chirality), signifies the process of creation of the conserved quantity, the chiral charge C, from the vacuum.

This process is induced by the applied parallel magnetic and electric fields and results from the node (Fermi point) in the energy spectrum. For the relativistic spectrum $E(\vec{k}) = ck$ of massless particles the Fermi point takes place at $\vec{k}_0 = 0$. Later we show that this Fermi point in the spectrum of the chiral fermions has the same topological charge as the node in the A-phase. That is why these nodes have the same cosequences on the dynamics of the system, irrespective of the origin of the nodes. In particular, the linear momentum creation from the vacuum of the A-phase in Eq. (6.15) under the time and space dependent texture, which plays the part of the electric $(\partial_t \hat{l})$ and magnetic $(\vec{\nabla} \times \hat{l})$ fields respectively, proves to be in one-to-one correspondence with the creation of the chiral charge in particle physics.

We begin first with the chiral fermions under a pure magnetic field, i.e. we consider here the anomalies in the A-phase related to the static texture $\vec{B} = k_F \vec{\nabla} \times \hat{l}$. The magnetic field $\vec{B}$ interacting with chiral fermions results in three main anomalies in the static properties of the A-phase in textures: i) nonzero density of the quasiparticle states; ii) anomalous term in the supercurrent in Eq. (6.14); and iii) nonanalytic behavior of the London energy.

6.7. Spectrum of the Chiral Fermions in Magnetic Field

The energy spectrum E,

$$\mathbf{H}\psi = E\psi \;,\; \mathbf{H} = \pm c\vec{\tau}\cdot\left(\hbar\frac{\vec{\nabla}}{i} - e\vec{A}\right)\;,\qquad (6.16)$$

of the Weyl electron in uniform magnetic field $\vec{B} = \vec{\nabla} \times \vec{A}$ along, say, z axis may be found by taking the square of both sides of Eq. (6.16):

$$\mathbf{H}^2\psi = E^2\psi \;,\; \mathbf{H}^2 = c^2\left(\hbar\frac{\vec{\nabla}}{i} - e\vec{A}\right)^2 - e\hbar c^2\vec{\tau}\cdot\vec{B}\;.\qquad (6.17)$$

This corresponds to the conv ational problem for a nonrelativistic electron in a homogeneous magneti field, whose energy levels depend on the momentum k_z along the magnetic field direction, on the number n of the Landau level and on the spin projection τ_z along the field:

$$E_n^2(k_z) = k_z^2 + \hbar c^2 B(|e|(2n+1) - \tau_z e) . \qquad (6.18)$$

The most important property of this spectrum (see Fig. 6.2) is that it contains an anomalous branch, $n = 0$, $\tau_z = +$ (for positive electric charge e), which crosses the zero energy level. Due to chirality of the particles this branch is asymmetric in k_z:

$$E_{\text{anomalous}}(k_z) = Cck_z , \qquad (6.19)$$

where C is the chirality of the Weyl electron. All other branches of the energy spectrum have a gap and are symmetric under the inversion of the momentum k_z.

The quasiparticles, which fill the negative energy levels of the anomalous branch without gap, form the conventional normal one-dimensional Fermi liquid with the Fermi point at $k_z = 0$. As distinct from the Fermi point in three-dimensional momentum space of the homogeneous A-phase state, this is the Fermi point in one-dimensional k_z space, which by its physical properties is equivalent to the conventional Fermi surface in 3-dimensional space. So the Fermi point in three-dimensional space gives rise to the Fermi point in one-dimensional space when the magnetic field is applied. All the abnormal behavior of the chiral fermions in the magnetic field is related to the existence of this anomalous branch and of the normal Fermi liquid of the fermions on this branch.

6.8. *Anomalous Branch and Nonzero Density of States in the $\hat{l}$ Texture*

The first consequence of the anomalous branch is the nonzero density of states at zero energy, which is a characteristic property of the normal Fermi liquid. This density of the quasiparticle states is proportional to the conventional density of states on the Landau level

$$N(0) = \frac{|e|B}{2\pi^2 \hbar^2 c} . \qquad (6.20)$$

$E_n(k_z)$

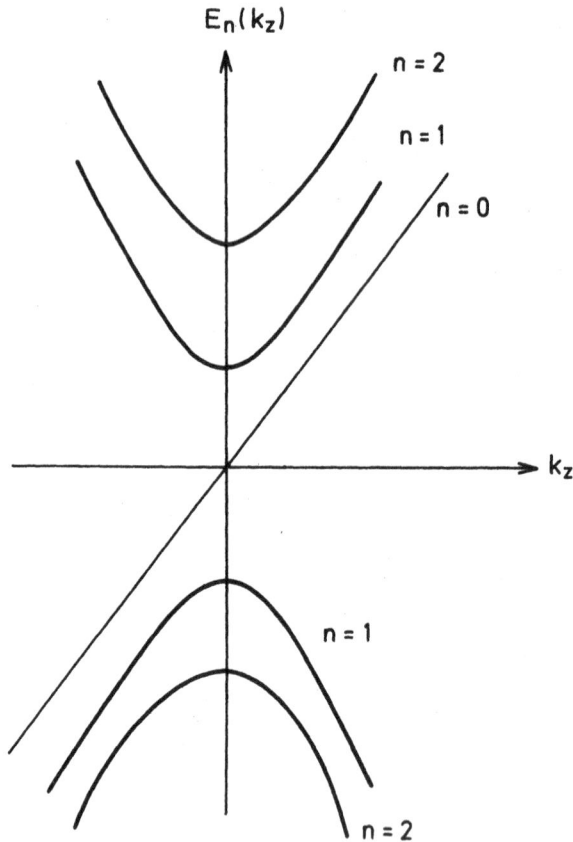

Fig. 6.2. Quasiparticle spectrum of the A-phase in the static $\hat{l}$ texture has the same structure as the spectrum of Weyl fermions in magnetic field, according to the analogy between $k_F \vec{\nabla} \times \hat{l}$ and magnetic field. n is the number of Landau level. For the chiral fermions there exists the anomalous asymmetric branch, which crosses zero level. The sign of asymmetry is defined by the chirality of the fermion, for a fermion with opposite chirality the anomalous branch is the mirror image of that in the figure. The gapless fermions on the anomalous branch form one-dimensional Fermi liquid with finite density of the quasiparticle states.

Now we can apply this to the A-phase. As distinct from the simple Weyl Hamiltonian for the chiral relativistic electron, the A-phase is very anisotropic, which is manifested by the anisotropic metric tensor in

Eqs. (5.15–19). This means that in the A-phase the gravity is well developed as compared with our Universe. Therefore Eq. (6.20) should be generalized to include the gravity field, i.e. the density of states should contain the dependence on the metric tensor. For fermions near the nodes, which obey the covariant and gauge invariant Weyl equation, the density of states is a covariant quantity under the general coordinate transformations. Therefore to apply Eq. (6.20) to the A-phase, one should rewrite it in the covariant form, which should be valid for all metric tensors including that of the A-phase state. The covariant and gauge invariant generalization of Eq. (6.20) is:

$$N(0) = \frac{|e| \sqrt{g}}{2\pi^2 \hbar^2} \sqrt{\frac{1}{2} g^{ij} g^{kl} F_{ik} F_{jl}} \ , \ F_{ik} = \nabla_i A_k - \nabla_k A_i \ , \qquad (6.21)$$

where g is the determinant of the metric tensor g_{ij}. In the isotropic case, when $g^{ij} = c^2 \delta^{ij}$ and $\sqrt{g} = 1/c^3$, one returns back to Eq. (6.20). Note that the velocity of light never enters explicitly the physical quantities such as density of states, but only through the metric tensor. Therefore there is no problem with the A-phase where the "velocity of light" depends on the direction of propagation.

In the particular case of the A-phase with $|e| = 1$, $\vec{A} = k_F \hat{l}$ and $\sqrt{g} = 1/c_\parallel c_\perp^2$ one obtains

$$N(0) = \frac{k_F^2}{2\pi^2 \hbar^2 \Delta_A} |\hat{l} \times (\nabla \times \hat{l})| \ . \qquad (6.22)$$

This means that the $\hat{l}$ texture gives rise to a nonzero density of states and therefore to a linear temperature dependence of the heat capacity, $C(T) \sim N(0)T$, at low temperatures, $T/\Delta_A \ll 1$. Without the textures the quasiparticles contribution to the heat capacity, which also comes mostly from the quasiparticles in the vicinity of the gap nodes, is essentially smaller: $C(T) \sim C(T_c)(T/\Delta_A)^3$.

The nonzero density of states due to the texture also gives rise to nonzero density ρ_n of the normal component at $T = 0$. This normal component is thus produced by the one-dimensional normal Fermi liquid on the anomalous branch.

6.9. *Zero Charge Effect and Nonanalyticity of the Magnetic Energy of the Vacuum*

Massless fermions have another distinguished property, which is important both for particle physics and for ^{3}He-A. Since they are gapless they are easily created from the vacuum, and that is why they can effectively screen the external electric charge. As a result the effective electric charge of the external particle is reduced according to

$$e_{\text{eff}}^2 = e^2/\epsilon(\omega) . \tag{6.23}$$

Here the dielectric constant of the vacuum, $\epsilon(\omega)$, is logarithmically divergent at low frequency:

$$\epsilon(\omega) = \frac{1}{3\pi} \frac{e^2}{\hbar c} \ln \frac{\Lambda^2}{\omega^2} , \tag{6.24}$$

leading to a complete screening at $\omega = 0$, which is known as zero charge effect. Here Λ is the ultraviolet cutoff parameter.

This effect is easily calculated considering the magnetic energy of the vacuum. The Lagrangian for the electromagnetic (em) field is

$$L_{\text{em}} = \frac{1}{8\pi} \epsilon(\omega)(\vec{B}^2 c^2 - \vec{E}^2) . \tag{6.25}$$

and when one considers a finite magnetic field $\vec{B}$, the frequency squared, ω^2, is limited by the energy difference between the Landau levels, which according to Eq. (6.18) is $\omega_{\text{cutoff}}^2 = 2|e|Bc^2/\hbar$. So the pure magnetic energy of the vacuum should be

$$E_{\text{magn}} = \frac{1}{24\pi^2} \frac{e^2}{\hbar c} \vec{B}^2 c^2 \ln \frac{\Lambda^2}{B} . \tag{6.26}$$

This may be obtained just by calculating the magnetic energy of the vacuum. The latter is the difference between the sums over the negative energies of the quasiparticles in vacuum with (Eq. (6.18)) and without magnetic field:

$$E_{\text{magn}} = E_{\text{vac}}(B) - E_{\text{vac}}(B = 0)$$

$$= N(0) \int dk_z \sum_{n, E<0} E_n(k_z) - \sum_{\vec{k}, E<0} E_{\vec{k}}(B = 0) = \frac{1}{24\pi^2} \frac{e^2}{\hbar c} c^2 \vec{B}^2 \ln \frac{\Lambda^2}{B} . \tag{6.27}$$

The logarithmic divergence at small B and therefore the nonanalytical behavior of the magnetic energy, which is equivalent to the zero-charge effect, result from the massless nature of the fermions.

6.10. *Nonanalytic London Energy of the* ^{3}He-A *Vacuum*

Again, to convert this result to the A-phase physics one should modify Eq. (6.26) for the magnetic energy to express it in the covariant form:

$$E_{\text{magn}} = \frac{e^2}{\hbar} \frac{\sqrt{g}}{24\pi^2} \frac{1}{2} g^{ij} g^{kl} F_{ik} F_{jl} \ln \frac{\Lambda^2}{\sqrt{\frac{1}{2} g^{ij} g^{kl} F_{ik} F_{jl}}} . \qquad (6.28)$$

Then for the particular case of the A-phase the following expression is obtained for the nonanalytic contribution to the London textural energy in Eq. (3.9b), which modifies the bend term with K_3:

$$E_{\text{nonanalytic}} = \frac{k_F^2 v_F}{24\pi^2 \hbar} (\hat{l} \times (\nabla \times \hat{l}))^2 \ln \frac{\Delta_A}{\hbar v_F \, | \, \hat{l} \times (\nabla \times \hat{l}) \, |} . \qquad (6.29)$$

Here we obtained only one of the three terms in the London textural energy in Eq. (3.9b), and found that the corresponding parameter K_3 logarithmically diverges at low temperature. The lower cutoff parameter in the logarithm is the maximum value among T^2/Δ_A, ω^2/Δ_A and $v_F \, | \, \hat{l} \times (\nabla \times \hat{l}) \, |$. The other terms cannot be calculated from only the analogy between ^{3}He-A and particle physics. This analogy is valid only in the vicinity of the nodes, therefore only those quantities coincide for the both quantum field theories which are defined by the quasiparticle states in the vicinity of nodes. These are, for example i) the density of states; ii) the nonanalytic magnetic energy, with the logarithm coming from the low-energy excitations; and iii) the production of the chiral charge from the vacuum under the effect of the chiral anomaly. They are described by the single metric tensor g^{ij}. The others, such as the first two terms in Eq. (3.9b) are defined by deep fermionic levels, where no analogy exists, since the general equation (5.11) for the quasiparticle Hamiltonian is far from being gauge invariant or generally covariant.

Also the anomalous current in Eq. (6.14) comes from the deep Fermi levels. This current results from the anomalous branch of the quasiparticle spectrum. Due to the asymmetry of this branch the filled negative Fermi levels have noncompensated momenta k_z. Summation over these momenta gives the net mass current in the presence of the $\hat{l}$ texture. This anomalous mass current cannot be calculated using the simple expression for the density of states in Eq. (6.20) as

$$N(0) \int k_z \, d(ck_z) \, ,$$

since it is valid only in the vicinity of the Fermi point, while the main contribution to the integral is given by the deep levels. More accurate calculations lead to Eq. (6.14).

6.11. *Chiral Anomaly and Nonconservation of the Vacuum Current*

Now we discuss the physical interpretation of the nonconservation of mass current in Eq. (6.15). From the point of view of many-fluid dynamics of ³He-A this is no puzzle. The low energy dynamics of liquid contains the dynamics of the coherent degrees of freedom (which we call superfluid motion, or vacuum motion, or dynamics of the degeneracy parameters, Goldstone fields or the gauge fields) which interact dynamically with the system of normal excitations − fermionic quasiparticles − which form the normal component of the liquid. The normal motion is frozen out at low temperatures in conventional superconductors, but survives in systems with gap nodes. Therefore it is not so strange that the momentum of the coherent vacuum motion can reversibly or irreversibly leak from the open windows of the gap nodes into the fermionic degrees of freedom. However, to understand the concrete character of the leakage of momentum, we need to continue our analogy with particle physics relating to the particles creation from the vacuum. We start first from the chiral anomaly. Let us now apply an electric field $\vec{E} \parallel \vec{B}$ to the vacuum of the chiral fermions in the presence of the magnetic field. Under the electric field the momentum of the particle increases according to the equation $\partial_t k_z = eE$. Due to this the right-handed particles, which fill the anomalous branch of the spectrum in the fermionic vacuum, start crossing the zero-energy level. This means that an excess of the positive chiral charge dynamically appears from the vacuum state. This is not compensated by the left-handed particles, since they fill the anomalous branch of the spectrum with the opposite momentum. On the contrary, under the electric field they just move into the Fermi sea leaving behind the holes, which are right-handed. Therefore both the anomalous branches produce right-handed particles from the vacuum under an electric field. This is what is called chiral anomaly.

The rate of production of the chiral charge ρ_C^{qp} of the quasiparticles is

$$\partial_t \rho_C^{qp} + \vec{\nabla} \cdot \vec{j}_C^{qp} = N(0)ecE = \frac{e^2}{2\pi^2\hbar^2}\vec{B}\cdot\vec{E} \ . \tag{6.30}$$

To apply this to ^{3}He-A, one should keep in mind the essential difference in the positions of the gap nodes in the momentum space in ^{3}He-A and in particle physics. For the chiral fermions in the Weyl equation the gap nodes for the left-handed particles and those for the right-handed particles occur at the same point, at $\vec{k} = 0$. In ^{3}He-A the nodes are separated in the momentum space: the left-handed particles have a node in their spectrum at $\vec{k} = k_F\hat{l}$, therefore each of these particle has linear momentum $\vec{k} = k_F\hat{l}$. While the right-handed particles have a node at $\vec{k} = -k_F\hat{l}$ and therefore their momentum is $\vec{k} = -k_F\hat{l}$. So the momentum and the chiral charge of the A-phase quasiparticles are related by $\vec{k} = -Ck_F\hat{l}$. This means that the creation of the chiral charge ρ_C is accompanied by the creation of the linear momentum of quasiparticles

$$\vec{j}_{qp} = -\rho_C^{qp} k_F\hat{l} \ . \tag{6.31}$$

From Eqs. (6.31–32) it follows that the rate of creation of the quasiparticle momentum, which is the same as the quasiparticle mass current, is

$$\partial_t \vec{j}_{qp} + \vec{\nabla}\cdot\Pi_{qp} = k_F\hat{l}\frac{e^2}{2\pi^2\hbar^2}\vec{B}\cdot\vec{E} = \frac{k_F^3}{2\pi^2\hbar^2}\hat{l}(\partial_t\hat{l}\cdot\vec{\nabla}\times\hat{l}) \ , \tag{6.32}$$

which is just the momentum loss during the superfluid vacuum motion, described by Eq. (6.15). So the total current, sum of vacuum current and the current of the quasiparticles $\vec{j} + \vec{j}_{qp}$, is conserved:

$$\partial_t(\vec{j}_{qp} + \vec{j}) + \vec{\nabla}\cdot(\Pi + \Pi_{qp}) = 0 \ . \tag{6.33}$$

Thus the chiral anomaly in the superfluid ³He-A is the reversible transfer of the momentum (chiral charge) from the coherent vacuum motion of the superfluid component of the liquid to the incoherent motion of the normal component, which consists of the quasiparticle excitations of the vacuum state. The transfer occurs through the topological windows (nodes, Fermi points) in the momentum space at which the quasiparticle energy spectrum crosses zero energy level. Physically the time and space dependent $\hat{l}$ texture of the vacuum motion (effective electric and magnetic fields) serves as a pump which pushes the Fermi sea of the fermionic vacuum through two orifices in the momentum space, say from the south to the north pole (see Fig. 6.3).

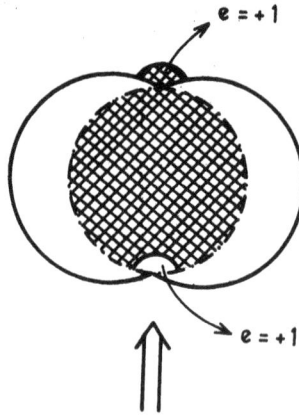

Fig. 6.3. Chiral anomaly in ³He-A: formation of the chiral charge $C = -e$ by the time and space dependent $\hat{l}$ texture. Under the effective electric and magnetic fields, $\vec{E} = k_F \partial_t \hat{l}$ and $\vec{B} = k_F \vec{\nabla} \times \hat{l}$, the Fermi sea is pumped through the gap nodes, producing quasiparticles at the north pole and quasiholes at the south pole, both with the same chirality. The chirality is related with the linear momentum of these fermions, $\vec{k} = -C k_F \hat{l}$, which gives rise to creation of the linear momentum of the quasiparticles from the coherent superfluid motion of the A-phase vacuum.

6.12. Dissipation in the Orbital Motion at Zero Temperature and Pair Creation in Electric Field in Particle Physics

The chiral anomaly — the creation of the conserved charge from the vacuum due to the dynamics of the $\hat{l}$ texture — is a reversible process at zero temperature. This is rather the flow of the quasiparticle energy levels through the Fermi points, than the transition of the particle from one level to another. So this process occurs without dissipation. Here we discuss another process related with the dynamics of the $\hat{l}$ texture, which is accompanied by the irreversible creation of the quasiparticles and corresponds to the pair creation from the vacuum under external electric field in particle physics. This process takes place when the effective electric field $\vec{E} = k_F \partial_t \hat{l}$, created by the time dependent texture, exceeds the effective magnetic field B. What occurs may be seen from Eq. (6.27) for the electromagnetic field Lagrangian generalized to include the electric field:

$$L_{em} = \frac{e^2}{\hbar c} \frac{1}{48\pi^2} (c^2 \vec{B}^2 - \vec{E}^2) \ln \frac{\Lambda^4}{c^2 \vec{B}^2 - \vec{E}^2} . \qquad (6.34)$$

The Lagrangian acquires the imaginary part when $E > cB$:

$$\text{Im}(L_{em}) = \frac{e^2}{\hbar c} \frac{1}{48\pi} (c^2 \vec{B}^2 - \vec{E}^2) , \quad E > cB , \qquad (6.35)$$

since the argument of the logarithm becomes negative. This means the dissipative process of the fermions creations from the vacuum, which is accompanied by the entropy production. This is the particular case of the pair production in electric field first calculated by Schwinger. In our case the fermionic quasiparticles have zero mass while Schwinger obtained the general result for the creation rate of massive particles: the probability of creation of the fermion-antifermion pairs per unit volume and unit time in a pure electric field is related with the following calculated imaginary part of the Lagrangian:

$$\text{Im}(L_{em})/\hbar = -\frac{1}{8\pi^3 \hbar} \frac{e^2}{\hbar c} \vec{E}^2 \sum_{n=1}^{\infty} \frac{1}{n^2} \exp\left(-\frac{n\pi M^2 c^3}{eE\hbar}\right) , \qquad (6.36)$$

where M is the mass of the fermion. For the massless particles, i.e., with $M = 0$, this transforms to Eq. (6.35).

Now one can apply the Lagrangian (6.34) to the $\hat{l}$ dynamics in the A-phase at $T = 0$. If the texture is homogeneous in space, one obtains the Lagrangian for the $\hat{l}$ field homogeneous dynamics from Eq. (6.34):

$$L_{electric} = -\frac{k_F^2}{48\pi^2 \hbar v_F} (\partial_t \hat{l})^2 \ln\left(-\frac{\Delta_A^2}{(\hbar \partial_t \hat{l})^2}\right) . \qquad (6.37)$$

This contains the imaginary part, which corresponds to the quasiparticle creation during the $\hat{l}$ dynamics if $(\partial_t \hat{l})^2 > v_F^2 (\vec{\nabla} \times \hat{l})^2$. So the $\hat{l}$ dynamics is dissipative even at zero temperature if the time dependence of the $\hat{l}$ field exceeds its spatial dependence. The pairs of fermions are created independently near each node: the fermion from, say, the Fermi sea near the north pole (i.e., with $\vec{k} \approx +k_F \hat{l}$, $k < k_F$) is excited into the state above the vacuum ($k > k_F$) but also in the vicinity of the same pole, $\vec{k} \approx +k_F \hat{l}$. So the total momentum is nearly conserved in this process and no net chiral charge appears, since this corresponds to creation of the particle with $e = +$ and the hole with $e = -$.

There are thus two main features, which distinguish the process of the pair creation in magnetic field from the chiral anomaly: 1) In this process the chirality $C = -e$ of the quasiparticles is conserved, and 2) this process is irreversible and therefore represents the dissipation of the energy of the vacuum motion into the quasiparticles energy.

6.13. *Wess-Zumino Action for the Orbital Dynamics*

Equation (6.37) in the Lagrangian describes the phenomena in the orbital dynamics related to the zero-charge effect in particle physics (real part of Eq. (6.37)) and the effect of the pair production in electric field (imaginary part of Eq. (6.37)). To this term in the Lagrangian, which of course should be generalized to include the spatial dependence of the order parameter (magnetic field $\vec{B}$), one should also add the term which describes the chiral anomaly. We obtain this term by considering the chiral anomaly equation (6.30).

This equation (6.30) may be rewritten as a continuity equation

$$\partial_\mu j_{C\mu} = 0 , \tag{6.38}$$

for the chiral current $j_{C\mu} = (\rho_C, \vec{j}_C)$ if one introduces the additional chiral current related to the vacuum motion. So the total chiral current is the sum of the quasiparticle and vacuum chiral currents:

$$j_{C\mu} = j_{C\mu}^{vac} + j_{C\mu}^{qp} , \tag{6.39a}$$

$$j_{C\mu}^{\text{vac}} = \frac{e^2}{2\pi^2\hbar^2}e^{\mu\nu\alpha\beta}A_\nu\partial_\alpha A_\beta \ . \tag{6.39b}$$

In the A-phase this corresponds to the conservation of the total (quasi-particles + vacuum) mass current in Eq. (6.33). The vacuum chiral current is to be obtained from the Lagrangian of the vacuum motion since $j_{C\mu}^{\text{vac}} = \delta L/\delta(eA_\mu)$. So one must find the term in the action for the gauge field A_μ, which gives Eq. (6.39b) for the vacuum current.

Here we encounter a new situation: the variation of the action is well defined:

$$\delta S_{\text{NWZ}} = \frac{e^3}{4\pi^2\hbar^2}e^{\mu\nu\alpha\beta}A_\nu\partial_\alpha A_\beta\delta A_\mu \ , \tag{6.40}$$

while the action itself cannot be constructed in local form. Such type of action was described by Novikov and Wess and Zumino. The Novikov-Wess-Zumino action, S_{NWZ}, has the topological origin, which often leads to the quantization of the physical parameters.

One way to write the NWZ action analytically is to add a new fictitious dimension, i.e. to consider the 5-dimensional space $x_\mu = (\tau, t, \vec{r})$, whose boundary coincides with the physical $(t, \vec{r})$ space:

$$S_{\text{NWZ}} = \frac{e^3}{12\pi^2\hbar^2}\int d^5x e^{\mu\nu\alpha\beta\gamma}A_\nu\partial_\alpha A_\beta\partial_\gamma A_\mu \ . \tag{6.41}$$

The variation of this action is the full derivative:

$$\delta S_{\text{NWZ}} = \frac{e^3}{4\pi^2\hbar^2}\int d^5x e^{\mu\nu\alpha\beta\gamma}\partial_\gamma(A_\nu\partial_\alpha A_\beta\delta A_\mu) \ , \tag{6.42}$$

which transforms to the integral over the boundary of the 5-dimensional space, i.e. over the physical space, giving rise to Eq. (6.40).

The corresponding NWZ action should exist in the A-phase. For the spatially homogeneous texture $\vec{A}(t) = k_F\hat{l}(t)$ this leads to the following type of topological action:

$$S_{\text{NWZ}}^{\text{anomalous}} = \frac{\hbar}{2}C_0\int d^3x\, dt\, d\tau\ \hat{l}\cdot\partial_t\hat{l}\times\partial_\tau\hat{l} \ , \quad C_0 = \frac{k_F^3}{3\pi^2\hbar^3} \ . \tag{6.43}$$

The variation of this action over $\delta\hat{l}$ leads to the elimination of the fifth coordinate τ, if one takes into account that $\delta\hat{l}\cdot\partial_t\hat{l}\times\partial_\tau\hat{l} = 0$ since $\hat{l}$ is the unit vector:

$$\delta S_{\text{NWZ}}^{\text{anomalous}} = \frac{\hbar}{2}C_0 \int d^3x \, dt \, d\tau \left(\delta \hat{l} \cdot \partial_t \hat{l} \times \partial_\tau \hat{l} + \hat{l} \cdot \partial_t \delta \hat{l} \times \partial_\tau \hat{l} + \hat{l} \cdot \partial_t \hat{l} \times \partial_\tau \delta \hat{l} \right)$$

$$= \frac{\hbar}{2}C_0 \int d^3x \, dt \, d\tau \left(\partial_t (\hat{l} \cdot \delta \hat{l} \times \partial_\tau \hat{l}) - \partial_\tau (\hat{l} \cdot \partial_t \hat{l} \times \delta \hat{l}) \right)$$

$$= -\frac{\hbar}{2}C_0 \int d^3x \, dt \, (\hat{l} \times \partial_t \hat{l}) \cdot \delta \hat{l} \, .$$

So one has

$$\frac{\delta S_{\text{NWZ}}^{\text{anomalous}}}{\delta \hat{l}} = -\frac{\hbar}{2}C_0 \hat{l} \times \partial_t \hat{l} \, . \tag{6.44}$$

The NWZ action comes from the gap nodes region in the quasiparticle spectrum, where the A-phase has the same properties as the vacuum of the chiral electrons in quantum electrodynamics. However a somewhat similar contribution to the action exists even in the anomaly-free A-phase. The case is that the dynamics of the variables, which is governed by the non-Abelian symmetry group, like the dynamics of the orbital momentum, can be easily expressed in the Hamiltonian form using the Poisson bracket scheme of the corresponding Lie algebra, while the Lagrangian formalism sometimes does not exist in terms of local variables. In order to obtain Eq. (6.7) for the anomaly-free vector $\hat{l}$ in the Lagrange formalism one should also introduce the nonlocal action. To make it easier we consider the case when the liquid density ρ may be taken as a constant in time. In this case one may write the total NWZ action for the A-phase, which includes both the chiral anomaly and the nonlocality of the action:

$$S_{\text{NWZ}} = \frac{\hbar}{2} \int d^3x \left(C_0 - \frac{\rho}{m_3} \right) dt \, d\tau \hat{l} \cdot \partial_t \hat{l} \times \partial_\tau \hat{l} \, , \tag{6.43a}$$

which leads to the following variation

$$\frac{\delta S_{\text{NWZ}}}{\delta \hat{l}} = -\frac{\hbar}{2} \left(\frac{\rho}{m_3} - C_0 \right) \hat{l} \times \partial_t \hat{l} \, . \tag{6.44a}$$

Here i) $C_0 = k_F^3/3\pi^2\hbar^3$ for the A-phase within the BCS theory in a weak coupling approximation; ii) $C_0 = 0$ for the anomaly-free A-phase-like liquid, and in the intermediate case iii) $C_0 = \frac{\tilde{k}^3}{3\pi^2\hbar^3}$, where $\tilde{k}\hat{l}$ is the position of the

Fermi point ($0 \leq \tilde{k} \leq k_F$). This topological term (6.43a) in action gives rise to the gap in the spectrum of the orbital waves which corresponds to the mass of the photon in particle physics (see the next subsection).

6.14. *Internal Angular Momentum of the A-phase and the Mass of Photon*

Now we can combine all the relevant terms in the Lagrangian (zero-charge-effect contribution + chiral anomaly contribution + conventional orbital momentum contribution) to obtain the dynamics equation of the $\hat{l}$ vector at $T = 0$:

$$\hat{l} \times \frac{\delta S}{\delta \hat{l}} = 0 , \quad S = S_{\text{em}} + S_{\text{NWZ}} . \tag{6.45}$$

This equation, according to Eq. (6.44a), has the form of conservation law for the orbital angular momentum of the Cooper pairs:

$$\partial_t \vec{L} = -\hat{l} \times \frac{\delta E}{\delta \hat{l}} . \tag{6.46}$$

It follows that the total orbital momentum $\vec{L}$ consists of two parts:

$$\vec{L} = \vec{L}_{\text{internal}} + \vec{L}_{\text{induced}} .$$

The internal angular momentum comes from the action S_{NWZ} and also contains two contributions:

$$\vec{L}_{\text{internal}} = (\hbar/2m_3)(\rho - m_3 C_0)\hat{l} . \tag{6.47}$$

The first contribution to the internal angular momentum $(\hbar/2m_3)\rho\hat{l}$ is just the momentum $\hbar\hat{l}$ per Cooper pair, this is what one must have in the anomaly-free A-phase. The second contribution, which in the real A-phase essentially compensates the effective value of this angular momentum of the vacuum, comes from the chiral anomaly action in Eq. (6.43). This nearly complete compensation corresponds to the reversible flow of the angular momentum from the vacuum to the fermionic quasiparticles, which occurs through the orifices at $\vec{k} = \pm k_F \hat{l}$ at the gap nodes.

The second part of the total angular momentum is the momentum induced by the motion of the $\hat{l}$ vector, which gives rise to the local angular velocity $\vec{\Omega} = \hat{l} \times \partial_t \hat{l}$. The induced angular momentum should be proportional to the angular velocity, like in a conventional mechanical system:

$$\vec{L}_{\text{induced}} = \chi_{\text{orbital}} \vec{\Omega} . \tag{6.48}$$

This term is obtained by variation of the Lagrangian (6.37) for electric field, which gives the following result for the orbital susceptibility:

$$\chi_{\text{orbital}} = \frac{k_F^2}{12\pi^2 \hbar v_F} \ln\left(-\frac{\Delta_A^2}{(\omega^2 - v_F^2(\vec{q} \cdot \hat{l})^2)}\right) . \tag{6.49}$$

Here we took as the lower cutoff parameter the frequency ω and the wave vector $\vec{q}$ of the $\hat{l}$ motion. The induced momentum was earlier neglected in Eq. (6.13), since for a very slow orbital dynamics the second time derivative which comes from the induced momentum is not important. But this second derivative term dominates when we consider the spectrum of the orbital waves, photons, which we discuss here.

The energy E in the rhs of Eq. (6.46) is the textural (magnetic) energy in Eqs. (6.28), (6.29). The other terms in the textural energy (3.9a) are neglected here since this logarithmically divergent term with K_3, which comes from the zero-charge effect, is dominating. So Eq. (6.46) is as follows:

$$\chi_{\text{orbital}} \hat{l} \times \partial_t^2 \hat{l} + L_{\text{internal}} \partial_t \hat{l} = K_3 \hat{l} \times (\hat{l} \cdot \vec{\nabla})^2 \hat{l} , \quad K_3 = \chi_{\text{orbital}} v_F^2 . \tag{6.50}$$

From this equation one may obtain the spectrum of the orbital waves (photons) at $T = 0$. At high frequency the term with internal angular momentum can be neglected and one has the linear spectrum of the electromagnetic waves:

$$\omega_{\text{photon}}^2(\vec{q}) = c_{\parallel}^2(\vec{q} \cdot \hat{l})^2 , \quad c_{\parallel} = v_F . \tag{6.51}$$

One can derive this spectrum in a phenomenological way directly from Eq. (5.16) for metric tensor, taking into account that $c_{\perp} \ll c_{\parallel}$. At lower frequencies two branches of this spectrum, which correspond to two polarizations of photon, are split due to internal angular momentum of the A-phase, which violates the time inversion symmetry, and therefore the spectrum depends on the spirality of the photon:

$$\omega^{\pm}_{\text{photon}}(\vec{q}) = \sqrt{c^2_{\parallel}(\vec{q} \cdot \hat{l})^2 + \left(\frac{L_{\text{internal}}}{2\chi_{\text{orbital}}}\right)^2} \pm \frac{L_{\text{internal}}}{2\chi_{\text{orbital}}} \ . \qquad (6.52)$$

The branch with spirality + has a gap which corresponds to the mass of the photon with this spirality:

$$\omega^{+}_{\text{photon}}(0) = \frac{L_{\text{internal}}}{\chi_{\text{orbital}}} \ , \qquad (6.53)$$

this gap is of order Δ^2_A/ϵ_F. The mass of the photon comes from the topological mass term in action, Eq. (6.43a). The orbital waves with negative spirality are massless, their spectrum is quadratic in the long wavelength limit:

$$\omega^{-}_{\text{photon}}(\vec{q}) \approx v^2_F(\vec{q} \cdot \hat{l})^2 \frac{\chi_{\text{orbital}}}{L_{\text{internal}}} \ . \qquad (6.54)$$

Such a splitting of the orbital waves into the branch with the gap and the branch with the quadratic spectrum resembles the spin-wave spectrum in ferromagnets and has the same origin. In ferromagnets the spin waves represent the oscillations of the spontaneous spin angular momentum in the same manner as the orbital waves in the A-phase represent the oscillations of the spontaneous orbital angular momemtum. In both cases the splitting of the collective modes results from the broken time inversion symmetry.

6.15. *Pair Creation by Accelerated Object and Unruh Effect*

Under the electric field in particle physics and under $\hat{l}$ dynamics in superfluid ^{3}He, in each elementary process of the dissipation two particles are created with opposite momenta. Different process of the irreversible pair creation is induced by the motion of the external object in particle physics vacuum or in the superfluid ^{3}He vacuum: the particles are created with equal momenta, and as distinct from the chiral anomaly this is a dissipative process. Let us compare this process in particle physics with that in the superfluid with a finite energy gap for quasiparticles, i.e. in the B-phase. In both cases the fermions are described by the Hamiltonian (5.8). If the object moves with constant velocity v, which is less than the Landau velocity in the superfluid ^{3}He and less than the light velocity in particle physics, then no dissipation takes place. The situation changes when the velocity depends on time, in this case the tunneling process of the particle creation takes place, and we estimate here the creation rate using the adiabatic perturbation theory.

If the external object moves along the trajectory $z = z_0(t)$ it produces a perturbation of the system, which depends on $z - z_0(t)$. The equation of motion for the quasiparticles is given as

$$i\partial_t\psi = \mathbf{H}_0\psi + \mathbf{H}_1(z - z_0(t))\psi \ , \qquad (6.55)$$

where $\mathbf{H}_1$ describes the perturbation which violates the translational symmetry due to the presence of the external object. If the wall of the container is moving in superfluids along the plane of the wall, we must assume that the surface of the wall is rough, otherwise the translational invariance is not broken. After the coordinate transformation $z = \tilde{z} + z_0(t)$ one obtains the equation with the time independent $\mathbf{H}_1$ but with the time-dependent Doppler term:

$$i(\partial_t - v(t)\partial_{\tilde{z}})\psi = \mathbf{H}_0\psi + \mathbf{H}_1(\tilde{z})\psi \ , \qquad (6.56)$$

where $v(t) = \partial_t z_0(t)$ is the time-dependent velocity of the external object.

The term $\mathbf{H}_1(\tilde{z})$ violates the translational invariance and provides the possibility of the nonconservation of momentum. It thus gives rise to the nonzero matrix element between the vacuum state with zero momentum and the state with momentum $2k_z$, which corresponds to the creation of the pair of fermionic quasiparticles with equal momenta k_z. Incidentally in the adiabatic approximation the precise form of this perturbation is irrelevant: in the tunneling process the most important contribution to the matrix element is given by the term with $v(t)$ which defines the leading exponent in the adiabatic approximation, while $\mathbf{H}_1(\tilde{z})$ is responsible only for the prefactor, which we neglect here. So in what follows we shall omit this perturbation.

The adiabatic approximation is valid when the velocity changes slowly as compared with the characterisic time of the Hamiltonian $\mathbf{H}$. In this regime one can use the adiabatic quasiparticle energy, which is obtained from Eq.(6.56) (without the perturbation term) by taking the Fourier transformation and considering the slow time argument in $v(t)$ as an external parameter:

$$E(\vec{k}, t) = \vec{k} \cdot \vec{v}(t) \pm E_{\vec{k}} , \qquad (6.57)$$

where

$$E_{\vec{k}} = \sqrt{m^2 c^4 + k^2 c^2} , \qquad (6.58)$$

in the relativistic system and

$$E_{\vec{k}} = \sqrt{\varepsilon_{\vec{k}}^2 + \Delta_B^2} , \qquad (6.59)$$

in the superfluid ^{3}He-B.

The amplitude of creation of the pair of fermionic particles with equal momenta corresponds to the particle transition from the filled negative energy level $E_-(\vec{k}_\perp, k_z, t)$ to the positive branch $E_+(\vec{k}_\perp, -k_z, t) = -E_-(\vec{k}_\perp, k_z, t)$ with the opposite k_z. In the adiabatic approximation this amplitude is given in terms of the classical action

$$S_+(\vec{k}, t) = \int^t dt_1 \, E_+(\vec{k}_\perp, -k_z, t_1) , \quad S_-(\vec{k}, t) = \int^t dt_1 \, E_-(\vec{k}_\perp, k_z, t_1) , \qquad (6.60)$$

in the following form:

$$A_{\vec{k}} = \int dt \, f(t) e^{-i(S_-(\vec{k}, t) - S_+(\vec{k}, t))/\hbar} = \int dt \, f(t) e^{-2iS_-(\vec{k}, t)/\hbar} . \qquad (6.61)$$

The prefactor $f(t)$ is unimportant in our approximation, since the leading contribution to the integral comes from the stationary point t_0 of the exponent: $E_-(\vec{k}, t_0) = 0$. Since we consider the velocity, which is less than the Landau critical velocity, the stationary point never occurs on the real time axis. So this is the tunneling process at which the stationary point t_0 is situated in the complex time plane. To find the main contribution one should choose such t_0 which is nearest to the real axis. So the transition probability is given by

$$w \sim \exp\left(-4 \, \mathrm{Im} \int^{t_0} (k_z v(t) - E_{\vec{k}}) dt\right) . \qquad (6.62)$$

This probability is essentially defined by the time-dependence of $v(t)$. Let us choose the following time-dependent velocity of the external body:

$$\vec{v}(t) = \hat{z}u \, \text{th} \, \frac{at}{u} \, , \tag{6.63}$$

where u is some limiting velocity, which is less or equal to the light velocity in particle physics and less than the Landau velocity in ³He. For small t this corresponds to the simple accelerated motion of the object with the acceleration a: $v(t) \approx at$. In the general case this function is the solution of the following equation:

$$\frac{\partial_t v}{1 - \frac{v^2}{u^2}} = a \, . \tag{6.64}$$

If one chooses $u = c$ in the relativistic theory one obtains that the parameter a means the constant acceleration of the external body in the frame moving with the object.

The stationary point t_0 is found from

$$\frac{k_z u}{E_{\vec{k}}} \, \text{th} \frac{at}{u} = 1 \, . \tag{6.65}$$

If $u < v_L$ (or $u \leq c$ in particle physics) then for all momenta $\vec{k}$ the solution for t_0 is complex:

$$\frac{at_0}{u} = \frac{\pi}{2}i + \frac{1}{2}\ln\frac{E_{\vec{k}} + k_z u}{E_{\vec{k}} - k_z u} \, . \tag{6.66}$$

This gives the following creation rate:

$$w \sim \exp -\frac{2\pi u(E_{\vec{k}} - k_z u)}{\hbar a}) \, . \tag{6.67}$$

So the particles, created at zero temperature by the object, accelerated according to Eq.(6.63) in the B-phase and in the vacuum of particle physics, are thermally distributed with the effective temperature

$$T_{\text{eff}} = \frac{\hbar a}{2\pi u} \tag{6.68}$$

both in the B-phase and in particle physics. For $u = c$ this corresponds to the Unruh temperature

$$T_{\text{U}} = \frac{\hbar a}{2\pi c} \, , \tag{6.69}$$

for the accelerated objects in particle physics: for the accelerated objects the vacuum looks like the thermal bath with $T = T_U$.

For condensed matter the chosen trajectory of the external body (wall of container, or vibrating wire) seems to be very artificial, though possible. In the experimental situation one should choose the oscillating velocity, like in the experiments with the wire vibrating in ^{3}He:

$$x(t) = \frac{u}{\Omega} \sin \Omega t . \tag{6.70}$$

Such oscillatory regime is also easier to construct in particle physics. The stationary point t_0 in this case is found from the following equation

$$\frac{k_z u}{E_{\vec{k}}} \cos \Omega t = 1 , \tag{6.71}$$

which for chosen $u < v_L = \Delta/k_F$ or $u \leq c$ has the imaginary value for all $\vec{k}$: $t_0 = i\tau_0$ with

$$\text{ch } \Omega\tau_0 = \frac{E_{\vec{k}}}{k_z u} . \tag{6.72}$$

This gives for the pair creation rate

$$w_{\vec{k}} \sim \exp \left(-4\frac{E_{\vec{k}}}{\Omega} \left[\ln \left(\frac{E_{\vec{k}}}{k_z u} + \sqrt{(\frac{E_{\vec{k}}}{k_z u})^2 - 1} \right) - \sqrt{1 - (\frac{k_z u}{E_{\vec{k}}})^2} \right] \right) . \tag{6.73}$$

For the small oscillation amplitude u this results in

$$w_{\vec{k}} \sim \exp \left(-4\frac{E_{\vec{k}}}{\Omega} \ln \frac{2E_{\vec{k}}}{ek_z u} \right) . \tag{6.74}$$

Within the logarithmic accuracy this can be interpreted as thermal radiation with the effective temperature:

$$T_{\text{eff}} = \frac{\Omega}{4\ln(v_L/u)} . \tag{6.75}$$

in the B-phase and

$$T_{\text{eff}} = \frac{\Omega}{4\ln(c/u)} . \tag{6.76}$$

in particle physics, where this corresponds to the multi-photon process of the pair creation.

In the B-phase the large Ω can be achieved in ultrasonic experiments since the superfluid velocity is oscillating in propagating sound. The superfluid velocity $\vec{v}_s$ has the same effect on the quasiparticle spectrum as the velocity of the moving object.

7

Topological Objects in Superfluid Phases of ^{3}He

7.1. *Quantum Number and Topological Charge*

The Bose and Fermi excitations of the superfluid vacuum of ^{3}He phases are described by quantum numbers (charges), whose conservation follows from the symmetry group of the vacuum state. Here we discuss other objects, whose stability is guaranteed by the conservation of another kind of charge, the topological invariant or topological charge, like the winding number. This topological invariant gives rise to existence of such particle-like objects in condensed matter, like dislocations in crystals, quantized vortices in superfluids and superconductors, domain walls in ordered magnets, disclinations in liquid crystals, solitons in systems with spin and charge density waves, etc.

The complicated order parameter and the hierarchy of interactions and length scales in superfluid phases of ^{3}He, such as the coherence length ξ, the dipole length ξ_d, the magnetic length, etc... give rise to the variety of topologically stable textures, many of them having been observed experimentally. Some of the topologically nontrivial textures of the A-phase have already been considered in Sec. 3 as examples of the inhomogeneous solutions of the London or/and Ginzburg-Landau equations: i) the object in the form of the surface: $\hat{d}$- soliton in Sec. 3.5; ii) linear objects: quantized vortices and disgyrations in the $\hat{l}$ field in Sec. 3.8; also in Sec. 3.1 we came across iii) the point defect on the surface of the container, the boojum, which should exist even in a ground state of the A-phase in the singly connected vessel, and iv) the monopole-like object with the singular tail.

We obtained these textures, $\hat{d}$-soliton, $\hat{l}$-disgyrations and quantized vortices, as exact solutions of the London or/and Ginzburg-Landau equations. However, this does not mean that such textures can be created in a real system. To guarantee the stability of these objects one should first investigate the stability of the corresponding solutions towards the small perturbations, which is already quite difficult without extended numerical calculations, since for most of the objects the structure of the order parameter cannot be found analytically. Then, the stability should be investigated towards large deformations of the order parameter, which is essentially more complicated. Therefore, taking into account a large variety of textures in superfluid ³He, finding the exact solutions as the first step in the investigation of all the possible inhomogeneous vacuum states proves to be the wrong way. Instead the following sequence of actions is usually taken.

7.2. *Topological and Symmetry Classification Schemes of Textures*

The theoretical investigation of textures in ³He and other condensed matter with spontaneously broken symmetry proceeds through three consecutive stages. At the first stage, the textures are distributed in large classes, defined by their distinct topological invariants, or the topological "charges". Due to conservation of topological charge, textures from a given topological class cannot be continuously transformed into a texture in another class, while the continuous deformation of the degeneracy parameter fields at which one texture transforms into another texture within the same class is allowed by topology. The homogeneous state has zero topological charge, therefore textures with nonzero charge are topologically stable: they cannot dissolve into the uniform vacuum state in a continuous manner, i.e. without creation of singularities of higher spatial dimension.

At this stage we do not need a detailed knowledge of the energy functional, neither do we need to solve any equation. The problem of the topological classification is solved by using purely topological methods. This classification is completely defined by the large symmetry group G and the residual symmetry group H of the vacuum state.

At the second stage, textures inside a given topological class are subdivided into symmetry classes — essentially in the same manner as in the classification of the bulk superfluid phases. This we already considered for the example of the AB interface, where two of the most symmetric states of the interface were obtained just from symmetry arguments. Again at this stage one does not need a detailed knowledge of the energy functional, everything is defined by some broken symmetry group and the residual symmetry group. In this case these groups characterize the symmetry of the inhomogeneous vacuum state: the vacuum state in the presence of the topological object. These symmetries are reduced as compared with the symmetry of the homogeneous vacuum due to the presence of a fixed line, fixed point or fixed plane, at which the topological object is concentrated, so they are defined mostly by the geometry of the defect.

At this second stage we find that there occurs an interplay between symmetry and topology: some symmetry classes are strongly prohibited in given topological classes, which automatically leads to the spontaneously broken symmetry inside some of the textures. A typical example: the space inversion symmetry P is necessarily broken in the A-phase continuous vortices. In the same manner as the states of condensed matter with different symmetry, the textures with different internal symmetries display different physical properties, such as spontaneous magnetic moment, spontaneous electric dipole moment, or spontaneous supercurrent. The phase transition between textures with different symmetries occurs in the same manner as the phase transition between different superfluid phases. Such type of transition has been observed in the rotating cryostat with ^{3}He-B; this is the transition between the quantized vortices with the same topological charge but different symmetry of the hard core.

Finally, only in the third stage is the numerical analysis of the textures carried out. Even at this stage it is not necessary to make a direct numerical computation of all possible textures. We are restricted by a given topological class and within this class by a given symmetry class. One then finds the texture with the minimal energy inside this symmetry class, and this texture automatically proves to be the exact solution of the London or Ginzburg-Landau equations, corresponding to the given topological charge and to given residual symmetry. Comparing the energies of the textures within the same topological class, but of different symmetry class, one then finds the possible transitions between textures displaying different residual symmetries.

7.3. Half-quantum Vortex and Combined Invariance

We apply this classification scheme to linear objects, such as quantized vortices, since they are investigated in detail both theoretically and experimentally. However let us first illustrate further the linear defects as exact solutions. Above we considered the singular lines in the orbital part of the degeneracy parameter (*dreibein*) which included the singularity in the $\hat{l}$ field, the $\hat{l}$ disgyrations, and in the θ_3 field, the quantized vortices. Now we proceed to the linear objects, related to the $\hat{d}$ field.

If one neglects the spin-orbit interaction, then the corresponding equation for the $\hat{d}$ field is

$$\hat{d} \times \frac{\delta F_{\text{grad}}^{\text{London}}}{\delta \hat{d}} = 0 \ . \tag{7.1}$$

Let us choose the $\hat{l}$ vector along z: $\hat{l} = \hat{z}$ and consider the $\hat{d}$ texture which depends on the transverse coordinates x and y. We can also choose the vector $\hat{d}$ concentrated in some plane, the plane may be defined by two unit orthogonal vectors $\hat{m}$ and $\hat{n}$. This large freedom of choice of the solutions illustrates the difficulty in constructing all possible inhomogeneous states without topological guide. So using our choice one has

$$\hat{d} = \hat{m} \cos \chi(x,y) + \hat{n} \sin \chi(x,y) \ . \tag{7.2}$$

Fortunately the equation for the angle χ proves to be very simple for a given choice:

$$\nabla^2 \chi = 0 \ , \tag{7.3}$$

and it has the following axisymmetric solutions:

$$\chi(x,y) = m_d(\phi - \phi_0) \ . \tag{7.4}$$

These solutions describe the singular lines, the disgyrations, in the $\hat{d}$ field with the integer winding number m_d; the angle χ is not well defined at the origin. These solutions hold only far from the origin, outside the hard core of order ξ.

One of the most interesting properties related to the spin disgyrations, is that the winding number m_d may be a half-integer. In such a case the angle χ changes by π around the disgyration, which means that the vector $\hat{d}$, when encircling the axis of disgyration, changes sign. This, however, may be compensated by the simultaneous rotation of the *dreibein* about the $\hat{l}$ vector by an angle π, which also changes the sign of the order parameter, since $(\hat{e}^{(1)} + i\hat{e}^{(2)}) \rightarrow -(\hat{e}^{(1)} + i\hat{e}^{(2)})$. The distribution of the A-phase order parameter around such singular line is

$$A_{\alpha i}(x,y) = \Delta_A \hat{d}_\alpha \left(\hat{e}^{(1)} + i\hat{e}^{(2)}\right)_i$$

$$= \Delta_A(\hat{m}_\alpha \cos m_d\phi + \hat{n}_\alpha \sin m_d\phi)(\hat{x}_i + i\hat{y}_i)e^{im_\Phi\phi} . \qquad (7.5)$$

If both the disgyration winding number m_d for $\chi = m_d\phi$ and the vortex winding number m_Φ for $\theta_3 = m_\Phi\phi$ are half integers, then the order parameter field is continuous around the axis. These are the exact solutions of the London equations for the $\hat{d}$ and θ_3 fields. The hard core structure of such combined defects again requires solving the Ginzburg-Landau equations.

For the particular case $m_d = \pm 1/2$ and $m_\Phi = \pm 1/2$ the order parameter in Eq. (7.5) describes the combination of the half-wounded disgyration with the half-quantum vortex (Fig. 7.1). The superflow velocity according to Eqs. (3.22–23),

$$\vec{v}_s = \frac{1}{2}\left(\frac{\hbar}{2m_3}\right)\frac{\hat{\phi}}{\rho} , \qquad (7.6)$$

is circulating about the vortex axis with one-half of the circulation quantum κ:

$$\oint d\vec{r} \cdot \vec{v}_s = \frac{1}{2}\left(\frac{h}{2m_3}\right) = \frac{1}{2}\kappa . \qquad (7.7)$$

The existence of such a combined object is the direct consequence of the discrete combined symmetry Z_2^{combined} (or in other notations P_1 symmetry) of the A-phase, which in an amazing way couples the spin part $\hat{d}$ of the

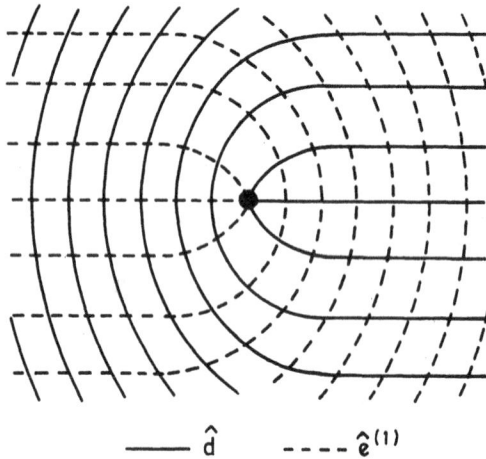

$$\underline{\hspace{1.5cm}}\ \hat{d} \qquad \text{-----}\ \hat{e}^{(1)}$$

Fig. 7.1. Half-quantum vortex in the A-phase. The $\hat{d}$ field (solid lines) performs the π winding around the vortex axis. The change of the sign of the order parameter is compensated by the π winding of the orbital *dreibein*; the dashed line represents the vector $\hat{e}^{(1)}$ of the *dreibein*: $\hat{e}^{(1)} = \hat{y}\cos\frac{\phi}{2} + \hat{x}\sin\frac{\phi}{2}$.

degeneracy parameter with the field of the *dreibein* $\hat{e}^{(1)} + i\hat{e}^{(2)}$. The coupling between these fields is purely topological, and is not energetical, since the equations for χ and θ_3 are completely independent: $\nabla^2\chi = 0$ and $\nabla^2\theta_3 = 0$. This coupling defines the common matching rule for χ and θ_3 when one goes around the singular line.

7.4. *Topological Classification of the Linear Defects*

So in the A-phase we have a variety of linear defects with hard core, disgyrations in the $\hat{l}$ and $\hat{d}$ fields, quantized vortices and numerous combined defects, combinations of the disgyrations with each other and with vortices. Now we must find out which of them are really stable, and which of them can be continuously transformed into each other.

This problem can be solved by pure topological methods. The degeneracy parameters $f(\vec{r})$ in a texture continuously depend on the coordinates everywhere except some isolated points, lines or surfaces. In this section, we consider the topological classification of the linear objects in superfluid ³He-

A, the topologically stable lines, at which the degeneracy parameters are not well-defined, i.e. linear defects. The spatial variation of $f(\vec{r})$ serves to define a continuous mapping of certain parts of the coordinate space X ($\vec{r} \in X$), where the degeneracy parameter is continuous, into the space $R = G/H$ of the degeneracy parameters ($f \in R$).

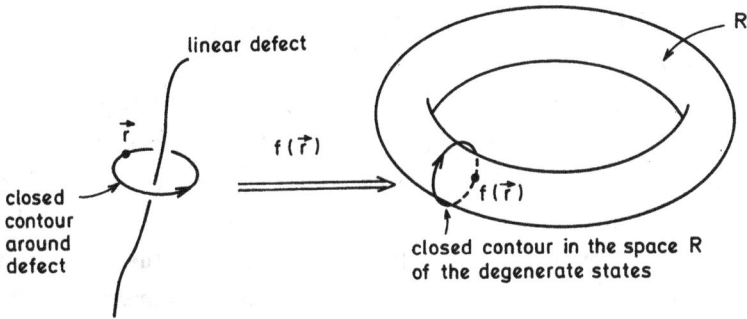

Fig. 7.2. Linear defects are described by the classes of continuous mapping of the contour around a defect into closed contours in the space R of the degenerate states. The mapping is realized by the distribution of the degeneracy parameter field $f(\vec{r})$ around a defect: each point $\vec{r}$ on the real space contour gives a point $f(\vec{r})$ in the space R.

In the case of linear defects the relevant manifold, whose mapping onto R gives all the information concerning the topological stability of the linear defect, is a closed contour around the defect line. The functions $f(\vec{r})$ transform this real space contour into a closed contour in the manifold R of internal states (Fig. 7.2). The closed contours in the compact space R may be distributed into classes. Within a given class one contour can be continuously transformed into another, while no such continuous deformation exists which can transform the contour from one class into the contour of another. Typical examples of the contours of different classes are the contours on the circumference $R = S^1$. The classes of contours are characterized here by the winding number N of the contour along the circumference. No continuous deformation between the contours with different N is possible.

It is known that the classes of closed contours in the compact space

R form a mathematical group, which is called the first homotopy group $\pi_1(R)$. Each class corresponds to some element of this group. The class of contractable contours corresponds to the unit element of this group. For the circumference S^1 this is the Abelian group Z of all integers N. For this example one can easily realize what is the group product of two closed contours: the contour which twice embraces the circumference in the positive direction is the product of two contours with unit winding number. This is the illustration of the arithmetic summation law for the group of integers: $1 + 1 = 2$.

So it follows that any possible linear defect in the degeneracy parameter field is thus described by some element of the first homotopy group $\pi_1(R)$. To identify the element of the homotopy group, which describes a given defect with given distribution of the degeneracy parameter fields $f(\vec{r})$, one should consider an arbitrary closed contour about the defect in real space X. Then the dependence $f(\vec{r})$ of the degeneracy parameter will trace the corresponding closed contour in the manifold R, and one must identify the class of this contour. As a result all the defects are distributed into topological classes, described by the elements of the homotopy group, $\pi_1(R)$. All the processes of fission and fusion of the defects are governed by the summation law within this group.

In the superfluid ⁴He, the manifold $R_{He-4} = U(1) = S^1$, therefore each linear defect is characterized by the integer $N = m_\Phi$, which is the phase Φ winding number, or the number of circulation quanta in quantized vortices. This topological charge is conserved for any continuous deformation, and the processes of fusion and fission of the vortices are regulated by a simple arithmetic summation law for this charge. In a formal description the homotopy group for the linear defects in the superfluid ⁴He is the group Z of integers:

$$\pi_1(R_{He-4}) = \pi_1(S^1) = Z \ .$$

7.5. Topology of Linear Defects in the A-phase

The manifold of the internal states R for the A-phase is the five-dimensional manifold

$$R_A = (S^2 \times SO_3)/Z_2 \,, \tag{7.8}$$

where S^2 is the sphere of the spin vector $\hat{d}$, and SO_3 is the space of the possible orientations of the *dreibein* $\hat{e}^{(1)}$, $\hat{e}^{(2)}$ and $\hat{l}$. The factorization of these spaces by the discrete group Z_2 reflects the discrete combined spin-gauge symmetry P_1 of the A-phase: the A-phase state does not change if the spin rotation which transforms $\hat{d}$ into $-\hat{d}$, is accompanied by the gauge transformation by π. The first homotopy group of this space

$$\pi_1(R_A) = Z_4 \,, \tag{7.9}$$

contains only four elements, i.e. there are only 4 topologically nonequivalent closed contours in the space R_A, and they form the cyclic group Z_4. This means the existence of four different topological classes of linear defects, to be identified by the numbers $N = 0, +1/2, -1/2$, and $1 = -1$, which thus does not coincide with the winding number m_Φ, as it takes place in ^{4}He.

The class with $N = 0$ is the class of the topologically trivial textures, which contains also the texture without singularities. It means that the linear defect of this class can continuously decay into a state without singularity. The topological charges N obey the summation laws by modulo 2, corresponding to the cyclic group Z_4, e.g. $1/2 + 1/2 = 1$, $1 + 1 = 2 = 0$.

The linear defects discussed above, the disgyrations and vortices, with the winding numbers m_l, m_d and m_Φ for $\hat{l}$, $\hat{d}$, and θ_3, are distributed among these four classes in the following way. i) The defects with even $m_l + m_\Phi$ and with any integer m_d belong to the trivial class $N = 0$, so the continuous unwinding of these defects is possible. It means that they can lose their hard core and transform into the nonsingular texture without any singularity, i.e., in such a way that the degeneracy parameters of the A-phase could be everywhere well-defined. So all the spin disgyrations with any integer m_d as well as doubly quantized vortices are topologically unstable. The unwinding does not necessarily lead to the homogeneous state, since besides the homotopy group $\pi_1(R)$ there can exist other topological invariants which classify the nonsingular textures.

ii) Defects with odd $m_l + m_\Phi$ belong to the class $N = 1$, this means that the radial and tangential disgyrations with $m_l = 1$, discussed in Sec. 7.3, can be continuously transformed into singly quantized vortices with $m_\Phi = \pm 1$, and vice versa. Which one of these textures is realized will depend on the energetics of the textures, which is defined in particular by the parameters in the London energy in Eq. (3.9). Usually the radial disgyration is more preferable, if the spin-orbital coupling is neglected. According to the summation rule $1 + 1 = 2 = 0$, the fusion of any two of these defects leads to their annihilation with formation of a nonsingular state.

iii) The half-quantum vortices, which have $m_\Phi = \pm 1/2$ and $m_d = \pm 1/2$, belong to the remaining two classes. The topological charge of these defects is defined completely by m_Φ, i.e. $N = m_\Phi = \pm 1/2$, while the sign of m_d is unimportant. This is because the difference in m_d between $1/2$ and $-1/2$ is an integer, and we know that any integer m_d is equivalent to zero.

7.6. Unwinding of the Singularity in the Doubly Quantized Vortex

The summation rules by modulo two may be checked by continuous deformation of the degeneracy parameter field. To illustrate the summation rule $1 + 1 = 2 = 0$, let us construct an explicit deformation which transforms the singular distribution of the degeneracy parameter in the doubly quantized vortex, Eq. (3.26a) with $m_\Phi = \pm 2$, into a nonsingular texture. Let us choose for simplicity $m_\Phi = -2$ vortex in Eq. (3.26a), in this case the deformation can be chosen as axisymmetric. So an initial texture is:

$$\hat{d} = \hat{z} , \quad \hat{e}^{(1)} + i\hat{e}^{(2)} = (\hat{x} + i\hat{y})e^{-2i\phi} = (\hat{\rho} + i\hat{\phi})\,e^{-i\phi} , \qquad (7.10)$$

with singularity on the axis, where the degeneracy parameters are not defined. In the corresponding axisymmetric deformation of the *dreibein*, which leads to the unwinding of the singularity, we shall continuously change the polar angle η of the $\hat{l}$ vector, which was initially zero in the whole texture, and we shall keep this value at infinity to conserve the asymptotic of the doubly quantized vortex. Our aim is not to destroy the vortex, but to remove the singularity on the vortex axis. So one has:

$$\hat{d} = \hat{z} , \quad \hat{l} = \hat{z} \cos\,\eta(\rho, t) + \hat{\rho}\sin\,\eta(\rho, t) ,$$

$$\hat{e}^{(1)} + i\hat{e}^{(2)} = \left(-\hat{z}\sin\,\eta(\rho, t) + \hat{\rho}\cos\,\eta(\rho, t) + i\hat{\phi}\right)e^{-i\phi} , \qquad (7.11)$$

where $0 \leq t \leq 1$ is the parameter of deformation. At $t = 0$ one has $\eta(\rho, 0) = 0$, which corresponds to the singular distribution in Eq. (7.10), and in the following momenta of time let us take

$$\cos \eta(\rho, t) = 1 - 2t \exp\left(-\frac{\rho}{\rho_0}\right) . \tag{7.12}$$

The singularity on the axis persists for all $0 \leq t < 1$, but its weight smoothly decreases to zero at $t = 1$. At this final moment, $t = 1$, there is no singularity at all. On the vortex axis, at $\rho = 0$, one has $\eta(\rho = 0, t = 1) = \pi$ and the complex vector becomes well-defined:

$$(\hat{e}^{(1)} + i\hat{e}^{(2)})_{\rho=0,t=1} = (-\hat{\rho} + i\hat{\phi})e^{-i\phi} = -\hat{x} + i\hat{y} \ , \hat{l}_{\rho=0,t=1} = -\hat{z} \ , \tag{7.13}$$

with the $\hat{l}$ vector directed in the opposite way as compared with its asymptotic value at infinity. The asymptotic behavior of this texture at $\rho \gg \rho_0$, where $\eta \to 0$, has the structure of the doubly quantized vortex in Eq. (7.10), so we came to the so-called *doubly quantized continuous vortex*. It has a smooth vortex core of the size ρ_0 (see Fig. 7.3), within which the degeneracy parameter of the A-phase is everywhere well-defined, with a fountain-like distribution of the $\hat{l}$ field.

If there are no other interactions the size of the soft core ρ_0 may be of order of the size of the vessel, or of the order of intervortex distance if there are several vortices in the vessel. The situation changes under external magnetic field, which interacts with the $\hat{d}$ vector, which in turn is coupled with the $\hat{l}$ vector via small spin-orbit interaction. In this case the core size depends on the behavior of the $\hat{d}$ field, and is either of the dipole length ξ_d, if $\hat{d}$ does not follow $\hat{l}$, or of the order of magnetic length if $\hat{d}$ is aligned with $\hat{l}$. The topological transition between these two types of doubly quantized vortices will be discussed in Sec. 7.9.

The velocity distribution is also continuous everywhere within the soft core:

$$\vec{v}_s = -\frac{\hbar}{2m_3\rho}(1 + \cos \eta(\rho))\hat{\phi} \ , \tag{7.14}$$

since the divergence at $\rho = 0$ is compensated by the factor $(1 + \cos \eta(\rho))$ which is zero on the vortex axis. The continuous vorticity $\vec{\nabla} \times \vec{v}_s$ is concentrated within the soft core.

(a)

(b)

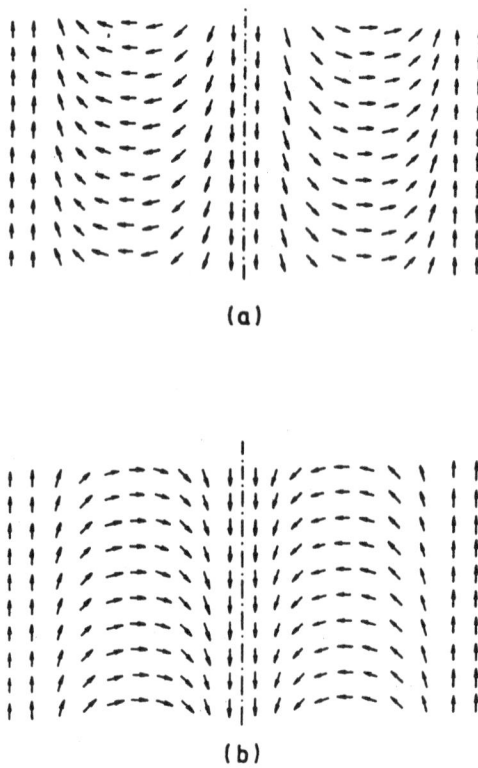

Fig. 7.3. Continuous vortex with the $m_{\Phi} = -2$ winding number may be obtained by smoothening of the vortex sheet of the singular vortex in Fig. 3.6b, which separates the peripheric A-phase state with $\hat{l} = \hat{z}$ from the A-phase state with $\hat{l} = -\hat{z}$ on the vortex axis. This flaring out of singularity occurs due to the $\hat{l}$ texture which continuously connects two orientations of $\hat{l}$. Two non-singular ways of unwinding of singularity, (a) and (b), are shown. These two nonsingular textures have the same energy and transform into each other under the symmetry operation P – space inversion. This means that the spatial parity P is spontaneously broken in the continuous vortex.

In the rotating cryostat these continuous vortices form an *Abrikosov* lattice, continuous periodic texture (see Fig. 7.4). These vortex textures were observed in the NMR experiments through the excitation of the spin waves localized in their soft cores. The topological transition between the

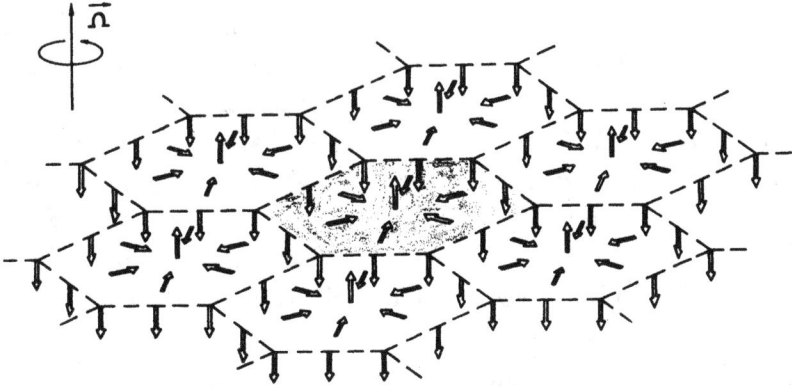

Fig. 7.4. Periodic texture of doubly quantized continuous vortices in the rotating vessel. Each elementary cell (one of which is darkened) comprises a vortex with $m_\Phi = 2$, so the circulation of the superfluid velocity around the elementary cell contains two elementary quanta. When the primitive vortex cell is swept, the $\hat{l}$ texture covers completely the unit sphere $\hat{l} \cdot \hat{l} = 1$, so the texture in the cell has a topological invariant in Eq. (7.20) equal to $\tilde{m}_l = 1$.

vortices has been observed in the ultrasonic experiments.

7.7. *Parity Breaking in the Continuous Vortex*

The symmetry requirement is often incompatible with topological constraints. In particular, the space parity P should be broken in continuous doubly quantized vortex textures. The $\hat{l}(\vec{r})$ texture transforms under the parity P transformation as

$$\mathbf{P}\hat{l}(\vec{r}) = \hat{l}(-\vec{r}) . \tag{7.15}$$

In particular for the $\hat{l}$ texture in Eq. (7.11)

$$\mathbf{P}\hat{l} = \mathbf{P}\left(\hat{z}\cos\ \eta(\rho,t) + \hat{\rho}\sin\ \eta(\rho,t)\right) = \hat{z}\cos\ \eta(\rho,t) - \hat{\rho}\sin\ \eta(\rho,t) . \tag{7.16}$$

Therefore the P symmetry, which requires

$$\hat{l}(\vec{r}) = \hat{l}(-\vec{r}) , \tag{7.17}$$

is satisfied in the initial singular vortex with $\eta = 0$ and is broken in the continuous vortex with $\eta \neq 0$. (Note that in superfluid ^{3}He, instead of the space parity P, which is already broken even in the homogeneous state, the combined parity-gauge symmetry $Pe^{i\pi}$ is conserved. However for physical quantities, which are gauge invariant, like $\hat{l}$ and $\vec{v}_s$, the symmetry $Pe^{i\pi}$ is indistinguishable from P. So for the gauge invariant physical variables the parity P is conserved in the homogeneous state, but broken in continuous vortices. As we shall see below this physical breaking of parity leads to new effects, such as spontaneous electric polarization, or/and spontaneous mass current along the vortex axis.)

So the spatial inversion symmetry P proves to be incompatible with the continuous distribution for the $\hat{l}$ field in the nonsingular $m_\Phi = -2$ vortex. Therefore the transition from singular to nonsingular vortex is accompanied by the spontaneously broken parity P in the soft core region of continuous vortex. As a consequence of the spontaneously broken discrete symmetry P, the vortex texture should be twofold degenerate. The second degenerate state is obtained through the parity transformation in Eq. (7.16). As a result there are two different states, see Fig. 7.3 , with the same energy, the same asymptotic behavior $\hat{l}(\infty) = \hat{z}$, and even with the same distribution of the superfluid velocity field, but with the mirror reflected distribution of the $\hat{l}$ field. These are :

$$\hat{l}_\pm(\vec{r}) = \hat{z}\cos\,\eta(\rho) \pm \hat{\rho}\sin\,\eta(\rho) \ . \tag{7.18}$$

The spontaneously broken parity proves to be the common property shared by vortices both in the A and B phases, which results from smoothening of the singularity in the hard core region. For the broken spatial parity in the core of the B-phase vortex see Sec. 8.

7.8. *Topology of the Continuous Textures. Second Homotopy Group*

We found that there exist several different textures belonging to the same topological class $N = 0$: singular and continuous textures, with even $m_\Phi + m_l$ at their asymptote. Further we consider the symmetry classification scheme for the texture, but here we note that there is another fine topological number, which characterizes the continuous vortex texture. This number arises if the $\hat{l}$ texture is fixed in the asymptotes, say, by action of the magnetic field and dipole forces, or if this texture is periodic, like in the Abrikosov vortex lattice.

This quantum number may be obtained if one integrates the Mermin-Ho relation over the soft core of the continuous vortex with even m_Φ winding number:

$$\frac{1}{2}\left(\frac{2m_3}{h}\right)\int_\sigma d\vec{s}\cdot\vec{\nabla}\times\vec{v}_s = \frac{1}{4\pi}\int_\sigma dx\,dy\left(\hat{l}\cdot\frac{\partial\hat{l}}{\partial x}\times\frac{\partial\hat{l}}{\partial y}\right). \qquad (7.19)$$

Here the surface σ crosses the soft core with the edge on the contour L outside the core, where the vorticity $\vec{\nabla}\times\vec{v}_s$ is absent. In the case of the periodic vortex lattice σ is the elementary cell of the lattice, and L the cell boundary. The left-hand side of this equation may be transformed to the integral over L

$$\frac{1}{2}\left(\frac{2m_3}{h}\right)\oint_L d\vec{r}\cdot\vec{v}_s = \frac{1}{2}m_\Phi,$$

which is the integer for vortices with even winding number m_Φ. So the right-hand side of Eq. (7.19) is also an integer, and this is the topological invariant

$$\tilde{m}_l = \frac{1}{4\pi}\int_\sigma dx\,dy\left(\hat{l}\cdot\frac{\partial\hat{l}}{\partial x}\times\frac{\partial\hat{l}}{\partial y}\right) \qquad (7.20)$$

of the $\hat{l}$ field. As distinct from the winding number m_l, which takes place only for the $\hat{l}$ textures constrained in a plane, this invariant $\tilde{m}_l$ describes the continuous mapping of the soft-core region σ onto the whole unit sphere S^2 of the vector $\hat{l}$. In this mapping produced by the $\hat{l}(\vec{r})$ field in the core, the $\hat{l}$ vector covers the unit sphere $\tilde{m}_l$ times, $\tilde{m}_l = 1$ for the doubly quantized vortex.

The quantum numbers m_l and $\tilde{m}_l$ have different topological nature. The first is related to the classes of mapping of the real space contours into contours in the space R of internal states. These classes form the π_1 homotopy group. The integer winding number m_l is a topologically conserved quantity, only if the $\hat{l}$ vector is restricted to a plane: in this case its space is the circumference S^1, which has topologically nontrivial closed contours. The situations where the $\hat{l}$ vector field is bound within the plane occurs very often in ^{3}He-A physics, for example in the presence of an external magnetic field, which thus changes the manifold R_A of the internal states of the A-phase.

The integer invariant $\tilde{m}_l$ is related to the classes of continuous mapping of the real space surface (σ) onto the surface in the space R. Here we consider the mapping of σ onto the unit sphere S^2 of the unit $\hat{l}$ vector. Such classes form the second, $\pi_2(R)$, homotopy group. The quantity $\tilde{m}_l$ shows how many times the $\hat{l}$ vector sweeps its unit sphere if $\vec{r}$ sweeps the surface σ. This $\tilde{m}_l$ is an integer and is a topologically conserved quantity only if the $\hat{l}$ field is somehow fixed outside the soft core region, which is the case in many physical situations. If $\hat{l}$ takes the same value for all points of the contour L, then the surface σ is topologically equivalent to the spherical surface, so one has the mapping $S^2 \rightarrow S^2$ which is always described by an integer $\tilde{m}_l$. That is why for the doubly quantized continuous vortex, with the $\hat{l}$ field being fixed at infinity, one has exactly $\tilde{m}_l = 1$: the $\hat{l}$ vectors completely sweeps the unit sphere once. Another physical example where $\tilde{m}_l$ is conserved is the periodic vortex lattice. In this case the elementary cell σ of the two-dimensional periodic lattice is equivalent to the two-dimensional torus, and the classes of continuous mapping of the two-dimensional torus onto the unit sphere S^2 of $\hat{l}$ are also characterized by the integer $\tilde{m}_l$. For the periodic vortex texture of doubly quantized vortices in Fig. 7.4, $\tilde{m}_l = 1$.

7.9. Topological Phase Transition in Continuous Vortices

One may construct a similar invariant for the $\hat{d}$ field as in Eq. (7.20) for the $\hat{l}$ field:

$$\tilde{m}_d = \frac{1}{4\pi} \int_\sigma dx\,dy \left(\hat{d} \cdot \frac{\partial \hat{d}}{\partial x} \times \frac{\partial \hat{d}}{\partial y} \right) . \tag{7.21}$$

This integer winding number shows how many times $\hat{d}$ covers its unit sphere $\hat{d} \cdot \hat{d} = 1$ when the elementary cell σ of the vortex lattice is swept. For the continuous vortex discussed in Eq. (7.11) (with $t = 1$), the $\hat{d}$ field was considered constant, and therefore the topological invariant $\tilde{m}_d = 0$ for such a vortex state. On the other hand, the spin-orbital interaction forces the $\hat{d}$ field to follow the $\hat{l}$ field, which means that if this small interaction prevails, then $\hat{d}$ has the same topological charge as $\hat{l}$, i.e., $\tilde{m}_d = \tilde{m}_l = 1$.

So, depending on the tiny spin-orbit interaction, two topologically different vortex textures can exist (Fig. 7.5), and one may expect the phase transition between these textures if one changes the external conditions which regulate the coupling of $\hat{d}$ and $\hat{l}$. It is important that such textural phase transition should be of the first order, since it is impossible to change continuously the integer topological invariant $\tilde{m}_d$ from 0 to 1; only an abrupt jump is possible.

An external parameter, which can cause topological transition in the A-phase vortices, is an external magnetic field. According to Eq. (2.14) the magnetic field $\vec{H}$ tends to orient the vector $\hat{d}$ in a plane transverse to $\vec{H}$ due to magnetic anisotropy of the A-phase. If the orientational effect of the magnetic field is larger than that of the spin-orbit coupling, i.e. if $(\chi_\perp - \chi_\parallel)H^2 \gg g_{so}$, then the vector $\hat{d}$ is decoupled from the $\hat{l}$ field (dipole unlocked) and is concentrated in the transverse plane. The topological invariant $\tilde{m}_d$ is exactly zero, $\tilde{m}_d = 0$, for the planar distribution of $\hat{d}$, since in this case $\hat{d}$ covers the line on its unit sphere and the area of the line is exactly zero (due to the topological stability of the invariant, it will be exactly zero even if there are some not very strong deviations of $\hat{d}$ from the plane). In the other extreme case of the small field $(\chi_\perp - \chi_\parallel)H^2 \ll g_{so}$, the field does not prevent the dipole locking of $\hat{d}$: an alignment of $\hat{d}$ along $\hat{l}$ takes place. In the latter case $\tilde{m}_d = \tilde{m}_l = 1$.

The first order transition between $\tilde{m}_d = 0$ and $\tilde{m}_d = 1$ should take place at some intermediate field where the magnetic anisotropy energy is comparable with the dipole energy, i.e. $(\chi_\perp - \chi_\parallel)H^2 \sim g_{so}$. This field is of order 20–50 Gauss. Experimentally this textural phase transition is observed (Fig. 7.6) by ultrasound transmission technique. The critical field at which topological transition occurs is found to be $H_c \approx 15$ Gauss. Hysteretic behavior with large metastability effects manifests the first order nature of this transition. The dipole-unlocked vortices with $\tilde{m}_d = 0$ are stable above H_c and metastable below H_c down to zero field. The dipole-locked vortices with $\tilde{m}_d = \tilde{m}_l = 1$ are stable below H_c and metastable above H_c down to the catastrophe field $H_{c1} \approx 40$ Gauss, at which they abruptly lose their stability.

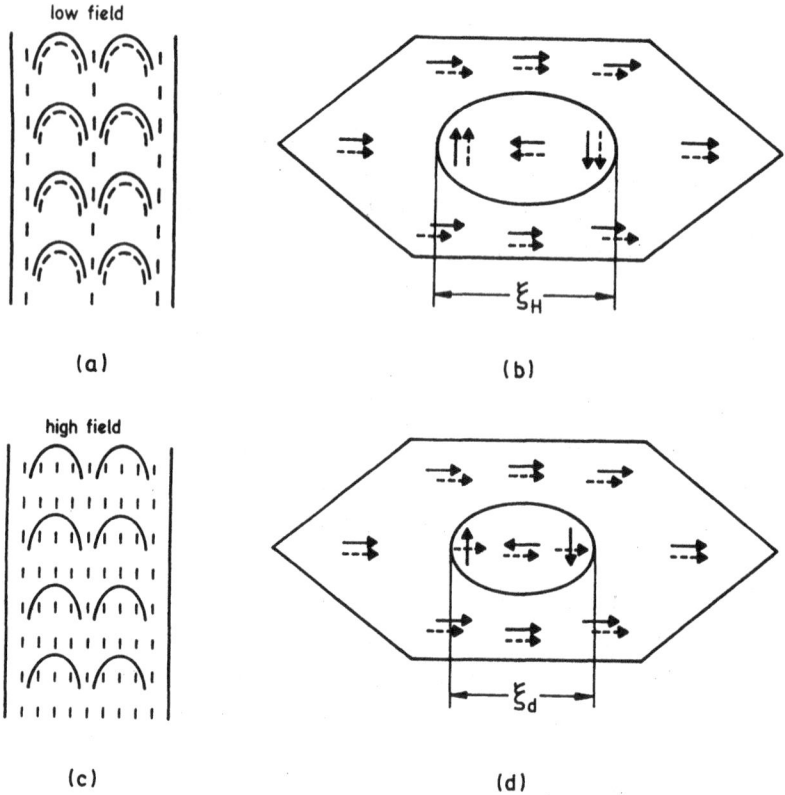

Fig. 7.5. Two topologically different continuous vortices in the A-phase. In (a) axisymmetric dipole-locked continuous vortex; the $\hat{d}$ field (dashed curves) follows the $\hat{l}$ field (solid curves) due to spin-orbit coupling, therefore $\hat{d}$ has the same topological invariant as the $\hat{l}$ field, $\tilde{m}_d = \tilde{m}_l = 1$. This dipole-locked vortex is stable at zero or low field. The primitive cell of the vortex lattice in low axial field is shown in (b) where ξ_H is the characteristic magnetic length, which defines the size of the soft core of dipole-locked vortex. Outside this core both $\hat{d}$ and $\hat{l}$ are in plane, therefore the axial symmetry is spontaneously broken in the vortices. In (c) dipole-unlocked axisymmetric vortex; the $\hat{d}$ field does not follow the $\hat{l}$ field and $\tilde{m}_d = 0$. The vortex with such topological invariant is stable at high field. The primitive cell of the dipole-unlocked vortex lattice in high field is shown in (d). Again axisymmetry is spontaneously broken.

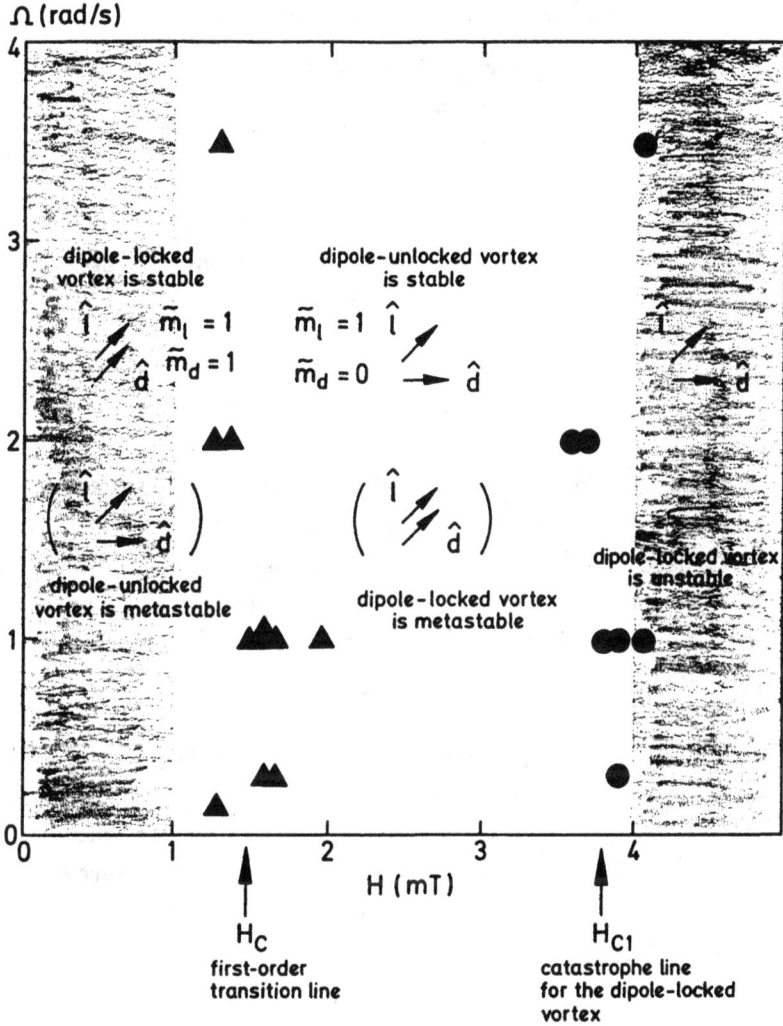

Fig. 7.6. Experimental phase diagram for first order topological phase transition between dipole-locked and dipole-unlocked continuous vortices in the $H - \Omega$ plane, where Ω is the angular velocity of rotation. Triangles mark the critical magnetic field H_c at which the transition is observed when H changes at given Ω. Circles represent the field above which the dipole-locked vortices are unstable.

7.10. *Hedgehog in the $\hat{d}$ Field. t'Hooft-Polyakov Monopole*

The process of changing the conserved quantity $\tilde{m}_d$ in the topological transition discussed above should occur with creation or emission of some object with the topological charge $\tilde{m}_d$. It is possible that the intermediate object, which carries the topological charge, is the pointlike defect in the $\hat{d}$ field, the hedgehog (Fig. 7.7a). The topological charge of the hedgehog is defined by the same expression as in Eq. (7.21), only the integration is taken over the surface embracing the point singularity:

$$\tilde{m}_d = \frac{1}{8\pi} \int_{\text{around hedgehog}} dS^i e_{ikl} \left(\hat{d} \cdot \frac{\partial \hat{d}}{\partial x_k} \times \frac{\partial \hat{d}}{\partial x_l} \right) . \tag{7.22}$$

This invariant shows the degree of the mapping of the surface S^2 embracing the point in real space onto the spherical surface S^2 of the vector $\hat{d}$. For the radial distribution of the spines of the hedgehog: $\hat{d} = \pm \hat{r}$ (here $\hat{r}$ is a unit vector of the spherical coordinate frame), and Eq. (7.22) gives $\tilde{m}_d = \pm 1$. The hedgehog with spines inward and therefore with $\tilde{m}_d = -1$ is shown in Fig. (7.7a).

In this scenario of the topological transition the topologically charged intermediate object, the hedgehog, is created at the edge of a vortex line, say, on the bottom wall of the container, and moves along the vortex axis to the other termination point of the vortex on the top wall, where it annihilates on the wall. Since its topological charge is just the difference of the integrals between the vortex cross sections above and below the hedgehog:

$$\int_{\text{around hedgehog}} = \int_{\text{above hedgehog}} - \int_{\text{below hedgehog}} , \tag{7.23}$$

the motion of the hedgehog thus changes the charge $\tilde{m}_d$ of the vortex (see Fig. 7.7).

According to the analogy between the quantum field theories in the A-phase and in particle physics, the $\hat{d}$ vector plays the part of the Higgs field, since it represents the quantization axis for the quasiparticle spin in the A-phase which is equivalent to the quantization axis for the weak isospin in particle physics. The hedgehog in the Higgs field represents the magnetic

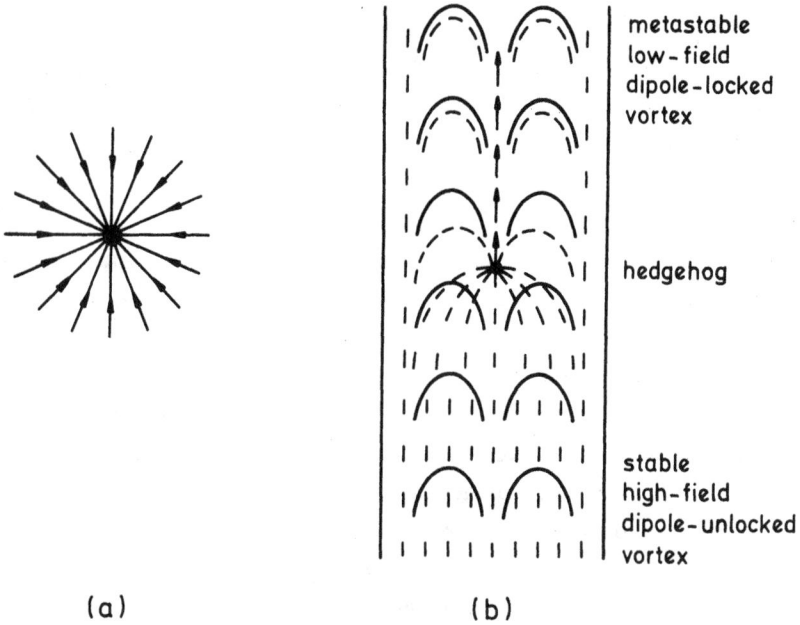

Fig. 7.7. a) The hedgehog in the $\hat{d}$ field, which is analogous to the t'Hooft-Polyakov magnetic monopole. b) The topological transition between vortex textures with different topological charge $\tilde{m}_d$, mediated by the monopole. At $H > H_c$ the dipole-locked vortex with $\tilde{m}_d = 1$ is metastable. It perishes when the monopole in the $\hat{d}$ field with $\tilde{m}_d = -1$ moves ahead and annihilates the topological invariant, leaving behind the stable dipole-unlocked vortex texture.

monopole, as was found by t'Hooft and Polyakov (as distinct from the Dirac magnetic monopole, which we shall discuss in the following subsection, the t'Hooft-Polyakov magnetic monopole has no string). So the quasiparticle spectrum near the A-phase hedgehog should be equivalent to the spectrum of the massless electron in the vicinity of the t'Hooft-Polyakov magnetic monopole.

7.11. *Hedgehog in the $\hat{l}$ Field. Dirac Monopole*

The point defects in the ordered matter are thus described by the second homotopy group $\pi_2(R)$ of the manifold of internal states, whose elements are classes of continuous mapping of the sphere S^2 onto R. A formal calculation of this group for the A-phase manifold R_A in Eq. (2.19) gives

$$\pi_2\left((S^2 \times SO_3)/Z_2\right) = \pi_2(S^2) + \pi_2(SO_3) = Z + 0 = Z \ . \qquad (7.24)$$

Only the sphere S^2 of the $\hat{d}$ vector contributes to this group, i.e. only the $\hat{d}$ field point defects can exist. The *dreibein* field with its manifold SO_3 cannot produce topologically stable point defects.

On the other hand, let us consider only the gauge invariant physical variables, $\hat{l}$ and $\vec{v}_s$, treating $\hat{e}^{(1)}$ and $\hat{e}^{(2)}$ as some hidden variables. The $\vec{v}_s$ field has no apparent topology, while the $\hat{l}$ field is restricted to be on the unit sphere $\hat{l} \cdot \hat{l} = 1$. In this description there are topologically stable point defects, hedgehogs, in the pure $\hat{l}$ field. How is this compatible with the absence of the stable point defects in the *dreibein* field? The answer is that the point defects in the $\hat{l}$ field is always accompanied by the vortex tail, the singular vortex line in the $\vec{v}_s$ field (or in the field of the hidden variables $\hat{e}^{(1)}$ and $\hat{e}^{(2)}$, which give rise to the velocity field of the vortex), terminating on the hedgehog (see Fig. 3.1). So there are no isolated point defects in the $\hat{l}$ field, the topological confinement always takes place between the hedgehog in the $\hat{l}$ field and the singular doubly quantized vortex. This vortex tail looks like a string, attached to the Dirac magnetic monopole. Below we shall show that the analogy is deeper, if one identifies the superfluid velocity $\vec{v}_s$ in the vicinity of the $\hat{l}$ hedgehog with the vector potential $\vec{A}$ of the electromagnetic field in the vicinity of the magnetic monopole.

To describe this combined defect, let us return to the singular vortex with $m_\Phi = -2$ and unwind only some section of the line, say the lower half-line in Fig. 7.8. Then one obtains the following distribution of the *dreibein* field in the spherical coordinate frame with unit vectors $\hat{r}$, $\hat{\theta}$, $\hat{\phi}$:

$$\hat{l} = \hat{r} \ , \quad \hat{e}^{(1)} + i\hat{e}^{(2)} = \left(\hat{\theta} + i\hat{\phi}\right) e^{-i\phi} \ . \qquad (7.25)$$

This structure has an $\hat{l}$ hedgehog with $\tilde{m}_l = 1$. On the lower half-axis $(z < 0)$, where $\hat{\theta} = -\hat{\rho}$ and $\hat{l} = -\hat{z}$, one has singularity-free *dreibein* in

Eq. (7.13). Singularity exists however on the upper half axis ($z > 0$): here $\hat{\theta} = \hat{\rho}$ and $\hat{l} = \hat{z}$, and one has the singular doubly quantized vortex in Eq. (7.10), attached to the hedgehog.

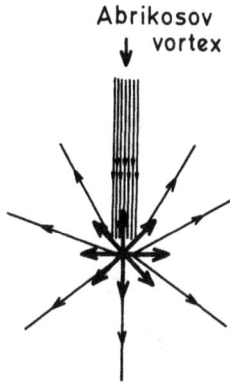

Fig. 7.8. Dirac magnetic monopole in the superconducting A-phase. The hedgehog in the $\hat{l}$ field (thick arrows) is the termination point of the Abrikosov vortex filament, which plays the part of a Dirac string. The magnetic flux, emanating from the magnetic pole (thin arrows show the direction of magnetic field) is supplied by the flux along the Abrikosov vortex. The magnetic charge of the monopole is quantized: $g = \tilde{m}_l \hbar c / 2e$.

The velocity field in this structure is given by Eq. (7.14) with $\eta(\vec{r}) = \theta$:

$$\vec{v}_s(\vec{r}) = -\frac{\hbar}{2m_3 r}\frac{1 + \cos\theta}{\sin\theta}\hat{\phi} \, , \qquad (7.26)$$

which looks just like the vector potential field in the Dirac magnetic monopole.

The similarity with the magnetic monopole increases if one considers the hypothetical electrically charged A-phase, say a superconductor with A-phase pairing. Here we use one of the important properties of superconductors — the Meissner effect: the magnetic field and therefore the electric current should vanish in the bulk of the superconductor. Therefore according to the London Equation (3.5) for the electric current in superconductors, the vector potential should compensate the superfluid velocity field in bulk liquid $\vec{A}(\vec{r}) = \frac{m_3}{e}\vec{v}_s$. Therefore outside of the hedgehog one has the $\vec{A}$ field in the form of the Dirac monopole:

$$\vec{A}(\vec{r}) = \frac{m_3 c}{e}\vec{v}_s = -\frac{\hbar c}{2er}\frac{1 + \cos\theta}{\sin\theta}\hat{\phi} , \qquad (7.27)$$

with the magnetic charge

$$g = \frac{\hbar c}{2e} . \qquad (7.28)$$

The magnetic field $\vec{B} = \vec{\nabla} \times \vec{A}$ is proportional to the vorticity $\vec{\nabla} \times \vec{v}_s$ and imitates the magnetic field of the magnetic charge g, which is radial and proportional to $1/r^2$:

$$\vec{B} = \frac{m_3 c}{e}\vec{\nabla} \times \vec{v}_s = g\frac{\hat{r}}{r^2} - 4\pi g\Theta(z)\delta_2(\vec{\rho})\hat{z} . \qquad (7.29)$$

The second term in the right-hand side of this equation describes the singular vorticity of the doubly quantized vortex, concentrated on the upper half-axis ($z > 0$). Therefore our $\hat{l}$ hedgehog is not an isolated magnetic monopole, but rather one pole of the magnetic dipole, i.e. the end of a thin and semi-infinite solenoid filament, represented by the singular vortex on the upper half axis. The magnetic flux $4\pi g$ emanating from the point defect in radial directions is exactly compensated by the magnetic flux through the filament. Nevertheless, similar to the Dirac and t'Hooft-Polyakov magnetic monopoles, we have a quantization of the magnetic charge: $g = \tilde{m}_l \, \hbar c/2e$. This is the topological quantization in terms of the topological charge $\tilde{m}_l$ of the $\hat{l}$ hedgehog and the elementary electric charge.

7.12. *Monopole and Boojum. Relative Homotopy Group*

The monopole-like object can in principle be stabilized in a vessel of spherical shape, where the boundary conditions require the radial distribution of $\hat{l}$ on the walls. According to Fig. 3.1 three different possibilities can occur. i) The monopole inside the vessel with the singular doubly quantized vortex emanating from the $\hat{l}$ hedgehog. ii) The same monopole, but with the vortex filament being split into two singular singly quantized vortices. iii) The monopole object is attracted by the wall of the container forming a point singularity on the wall, a boojum, without any singularity in the bulk liquid.

Which one of these textures should take place depends on the energetics of the textures, which is defined by the parameters in the London energy in Eq. (3.9). At very low temperatures the K_3 parameter of the term containing $\hat{l} \times (\vec{\nabla} \times \hat{l})$ is logarithmically divergent due to the zero charge effect in the fermionic vacuum (see Sec. 6). As distinct from other textures in the spherical vessel, the monopole texture $\hat{l} = \hat{r}$ has $\vec{\nabla} \times \hat{l} = 0$ and therefore does not contribute to the divergent energy term. Therefore one may expect that the monopole has the lowest energy at $T \to 0$. However under usual conditions (at finite T) the texture without singularities in the bulk liquid has less energy, since the vortex which terminates on the monopole also has logarithmical contribution to the energy according to Eq. (3.23): $F \sim R \ln(R/\xi)$, where R is the radius of the spherical vessel, while the energy of the singularity-free texture, $F \sim R$, has no logarithmic term (at finite T).

So it is more energetically favourable for the monopole to move to the wall. However, due to the boundary conditions on the $\hat{l}$ vector, the monopole does not disappear completely on the wall, but forms a topologically stable surface singularity, boojum. The topological stability of this point object is guaranteed by another homotopy group which takes into account the boundary conditions. Due to the boundary conditions, the freedom of the degeneracy parameter is restricted, so the manifold of the allowed internal states on the wall is reduced. Now instead of the space R_A in the bulk liquid, the manifold of states on the wall is

$$R_A(\text{wall}) = ((S^2 \times S^1)/Z_2) \times Z_2 . \qquad (7.30)$$

Here we take into account that the $\hat{l}$ vector is fixed on the wall, so instead of the SO_3 space of the *dreibein* only the circumference of the θ_3 angle is left. Additional Z_2 means that $\hat{l}$ may be either parallel or antiparallel to the normal to the wall.

To describe the classes of the topologically nonequivalent point defects on the surface, one should cover this defect by a hemisphere and map the hemisphere into the manifold of degenerate states R_A in the bulk liquid, while the edge of the hemisphere, the circumference S^1 on the surface of the wall, is mapped into the manifold $R_A(\text{wall})$. Such a mapping, which

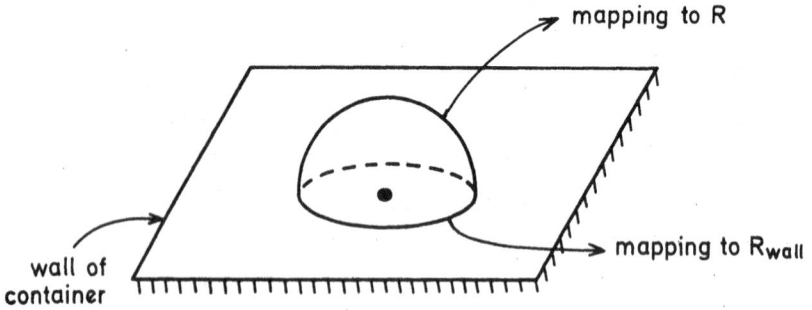

Fig. 7.9. Boojums are described by classes of continuous mapping of the hemisphere into the space R of all the degenerate states, which are possible in the bulk liquid, while the edge of hemisphere on the wall is mapped into the space R_{wall} of only those degeneracy states, which are allowed in the vicinity of the container walls by the boundary conditions.

takes into account the additional constraints on the boundary, is called *relative homotopy group*, and in the case of a hemisphere is denoted by $\pi_2(R, R(\text{wall}))$. The calculation of this group for the A-phase gives the group of integers $\pi_2(R_A, R_A(\text{wall})) = Z$. So the boojum is described by the integer number which just coincides with the topological charge $\tilde{m}_l$ of the monopole that came from the bulk liquid to form the boojum.

Related to the boojum is the circulation of the superflow around the boojum along the surface of the container. This circulation is also quantized with $m_\Phi = -2\tilde{m}_l$. Due to this the boojums play an important role in the rotating vessel. In the rotating vessel an equilibrium vortex lattice is always accompanied by the corresponding equilibrium number of the boojums on the top and bottom walls of the rotating cylindrical vessel (see Fig. 7.10). This comes from the following arguments. The circulation can be arbitrary in the bulk liquid due to continuous vorticity in the $\hat{l}$ texture, but it is strongly quantized on the surface, where the $\hat{l}$ vector is fixed and vorticity can be concentrated only at singular points. In the equilibrium rotating state the total circulation should be the same for all cross sections of the vessel, including the top and bottom walls, it is defined by the angular velocity Ω of the rotation of the vessel. Therefore the continuous vorticity of the doubly quantized vortex array should be transformed to the singular vorticity of the boojums on the top and bottom walls. This is the reason why a lot of boojums should exist in equilibrium rotating state.

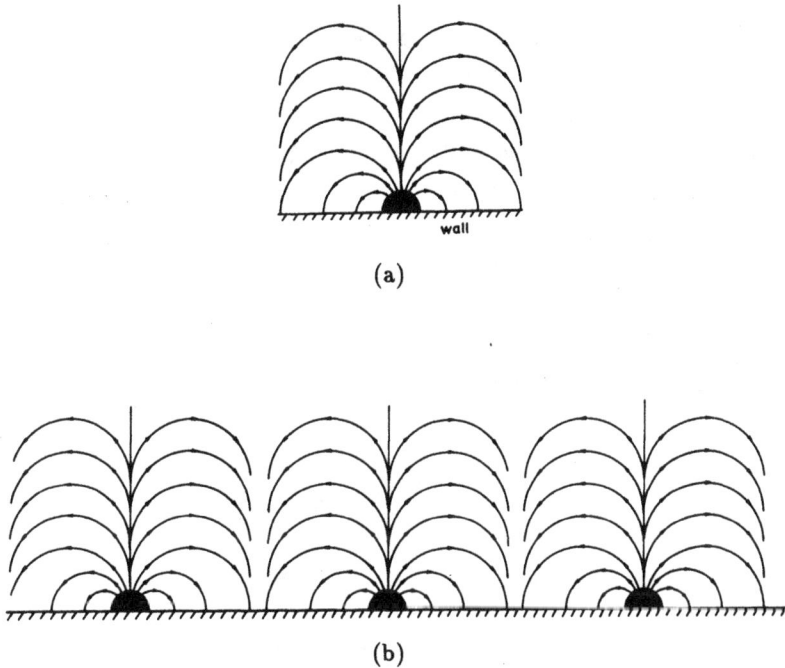

(a)

(b)

Fig. 7.10. a) A boojum is the end point of a continuous vortex. b) The lattice of boojums produced by the vortex lattice in a rotating vessel.

In principle, however, it is possible that in some cases, for example due to the strong pinning of the boojums on the container surface, it takes a long time to achieve the equilibrium state. So one may have metastable rotated state without boojums. This singularity-free state is somewhat unusual because the total circulation on the horizontal walls is exactly zero. Due to the conservation of circulation the total circulation should be exactly zero also for all cross sections of the rotating cylinder, and this is not compatible with the existence of the periodic array of nonsingular vortices, which produce the net Ω dependent circulation. The solution of this paradox is in the surface layer of the A-phase near the side walls of the container: in the singularity-free rotating state the surface layer consists of the compressed nonsingular vortices with the opposite circulation to compensate for the circulation of the bulk vortices.

Another type of the A-phase boojums (see Fig. 7.11) should exist on the AB interface, which provides tangential boundary conditions for the $\hat{l}$ vector (Sec. 3) as distinct from the normal boundary condition on the container wall. This boojum has topological charge $\tilde{m}_l = 1/2$ and represents the terminating point of the B-phase singly quantized vortex (string). In the electrically charged systems this boojum forms magnetic monopole with half of the elementary magnetic charge, possible in pure A-phase: $g = \frac{1}{2}\hbar c/2e$.

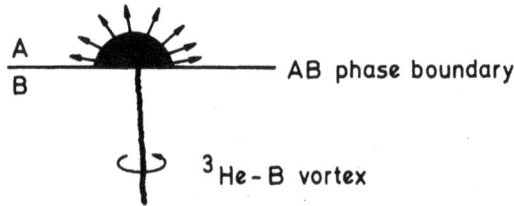

Fig. 7.11. Monopole-like object on the phase boundary between A and B phases. The role of the Dirac string is played by the singular B-phase vortex. The magnetic charge of this monopole is twice less than in the pure A-phase.

7.13. *Topology of Vortices and Disclinations in the B-phase*

The first homotopy group of the manifold of internal states for the B-phase is:

$$\pi_1(R_B) = \pi_1\left(S^1 \times SO_3\right) = \pi_1(S^1) + \pi_1(SO_3) = Z + Z_2 . \qquad (7.31)$$

It contains two independent topological charges N_1 and N_2, describing the linear objects in the B-phase.

i) The charge N_1 comes from the group $\pi_1(S^1) = Z$ of integers. It is the phase Φ winding number $N_1 = m_\Phi$ and describes the quantized vortices, which thus have the same topological properties as in superfluid ^{4}He, since in both cases the phase Φ is defined on the circumference S^1. The vortices with nonzero N_1 are topologically stable and have a singular core of the coherence length ξ size. Outside the core the asymptotic behavior of the phase is regulated by the analogous London equation for the phase:

$$\nabla^2 \Phi = 0 \ ,$$

as in ^{4}He, with the same solutions

$$\Phi = m_\Phi \phi \ , \tag{7.32}$$

and with superfluid velocity $\vec{v}_s = N_1 \hbar / 2m_3 \rho$. The singly quantized vortices have been observed in NMR experiments on rotating B-phase. However as distinct from the vortices in ^{4}He, many new physical effects are found which are related to the symmetry and structure of the hard core of the B-phase vortices, which we shall discuss in the next section.

ii) The charge N_2 comes from the homotopy group $\pi_1(SO_3) = Z_2$ of the SO_3 space of the orthogonal matrices $R_{\alpha i}$. This quantum number takes only two values 0 and 1 with the summation law $1 + 1 = 0$, which corresponds to the group Z_2. The linear defect of the nontrivial class $N_2 = 1$ is the disclination in the $R_{\alpha i}$ field.

To exemplify such defect let us express the matrix $R_{\alpha i}$ of rotations in terms of the direction of the rotation axis $\hat{n}$ and the angle θ of rotation about this axis:

$$R_{\alpha i}(\hat{n}, \theta) = \delta_{\alpha i} - (1 - \cos \theta)(\delta_{\alpha i} - \hat{n}_\alpha \hat{n}_i) - e_{\alpha i m} \hat{n}_m \sin \theta \ . \tag{7.33}$$

Then the space SO_3 may be considered as a space of all the nonequivalent vectors $\hat{n}\theta$. This is a solid sphere of radius π, since any rotation described by a vector outside of this sphere is equivalent to some rotation within the sphere. Also each two diametrically opposite points on the spherical surface, $\hat{n}\pi$ and $-\hat{n}\pi$, should be identified, since they correspond to the same rotation. The only topologically nontrivial closed contour Γ_1 in the space SO_3 just connects the diametrically opposite points (see Fig. 7.12b).

The stationary London equation for the degeneracy parameter $R_{\alpha i}$ is obtained as the time independent limit of Eq. (4.30a) of the spin dynamics:

$$e_{\alpha\beta\gamma} R_{\beta i} \frac{\delta F_{\text{grad}}^{\text{London}}}{\delta R_{\gamma i}} = 0 \ .$$

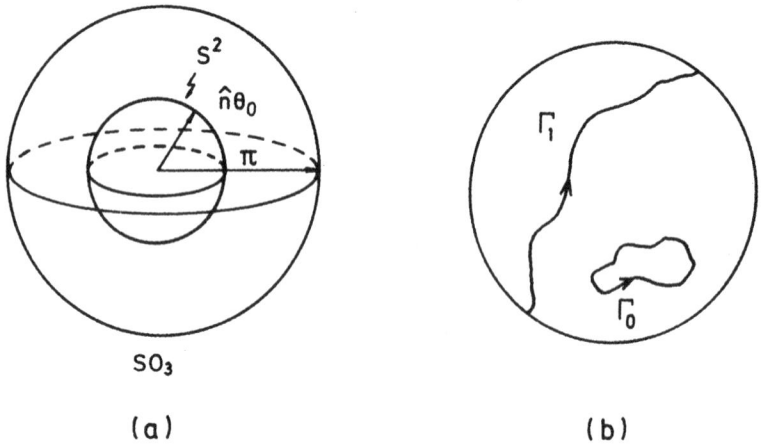

Fig. 7.12. (a) The SO_3 manifold is a solid sphere of radius π. Each point $\hat{n}\theta$ within the solid sphere corresponds to one solid rotation. Each two diametrically opposite points on the spherical surface, $\hat{n}\pi$ and $-\hat{n}\pi$, correspond to the same rotation and should be identified. The spin-orbital coupling reduces this space to the spherical surface of radius $104°$ (see below). (b) Contours Γ_0 and Γ_1 belong to two different possible classes of closed contours in the SO_3 manifold. The contour Γ_1 is closed since it connects two equivalent points. As distinct from Γ_0 the contour Γ_1 is topologically nontrivial since it cannot be contracted.

The simplest solution for the topologically nontrivial disclination is the 2π winding of the θ angle, $\theta = \phi$, at constant $\hat{n} = \hat{z}$. This solution gives the mapping of the closed contour around the disclination line into the vertical diameter of the solid sphere of the space SO_3 (Fig. 7.13), which is just the nontrivial contour Γ_1. This solution may be energetically unstable, but the instability will lead only to another solution of this topological class, which corresponds to some other contour of the same nontrivial class Γ_1 of closed contours in SO_3. In this sense the solution is topologically stable.

The linear objects, which have both topological charges, are also possible. These are combined defects. Especially interesting is the combination of the singly quantized vortex and disclination, $m_\Phi = 1$ and $N_2 = 1$, which was recently observed in NMR experiments. This defect was identified due to the spin soliton terminating on this line as a result of spin-orbital coupling.

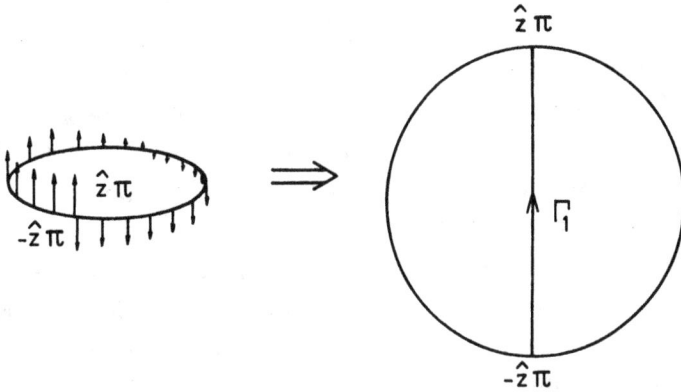

Fig. 7.13. Distribution of the vector $\hat{n}\theta$ on the closed contour γ around the simplest disclination with $N_2 = 1$. This distribution gives the mapping of γ to the vertical diameter in the space SO_3, which belongs to the topological nontrivial class Γ_1 of closed contours. Therefore this disclination is topologically stable.

7.14. *Soliton Terminating on the Disclination in the B-phase*

As in the case of the A-phase (see Sec. 3.7), the spin soliton in the B-phase arises due to a small spin-orbit interaction, which partially fixes the degeneracy parameter outside the soliton body.

The spin-orbital interaction, which couples the spin and orbital indices of the degeneracy parameter, has the following general form, which depends only on θ:

$$F_{\text{so}} = g_{\text{so}}^1 R_{\alpha\alpha} R_{\beta\beta} + g_{\text{so}}^2 R_{\alpha\beta} R_{\beta\alpha} = g_{\text{so}}^1 (1 + 2\cos\theta)^2 + 2g_{\text{so}}^2 (\cos^2\theta - \sin^2\theta) .$$

$$(7.34)$$

So, what is fixed in the B-phase degeneracy parameter at distances larger than the dipole length ξ_d, is the angle θ of the rotation, while the axis $\hat{n}$ of rotations still remains the Goldstone variable, since it rotates under the symmetry operation $SO_3^{(J)}$, which is the symmetry of the physical laws even in the presence of spin-orbital coupling. For the parameters $g_{\text{so}}^1 = g_{\text{so}}^2 > 0$ in the B-phase this expression is minimal at the "magic" angle

$$\theta_0 = \arccos\left(-\frac{1}{4}\right) \simeq 104° .$$

$$(7.35)$$

This equilibrium angle is manifested in various NMR experiments.

As a result of the spin-orbit coupling the space SO_3 of the internal states is reduced to the spherical surface S^2 for the vector $\hat{n}\theta_0$ within the solid sphere (see Fig. 7.12a). It is important that the structure of the disclination is in contradiction with this requirement: topologically stable disclination is defined by the small distance topology at the distances $\rho \ll \xi_d$, where the spin-orbital coupling can be neglected. This disclination is incompatible with the large distance topology at $\rho \gg \xi_d$. The large distance topology, which takes into account the spin-orbital coupling, does not support the existence of the topologically stable linear defects since $\pi_1(S^2) = 0$. From Fig. 7.14b it follows that the nontrivial (noncontractible) closed contour Γ_1 in the SO_3 space, which corresponds to the topologically stable disclination, cannot be restricted to the sphere S^2 of $\theta_0\hat{n}$ within this space.

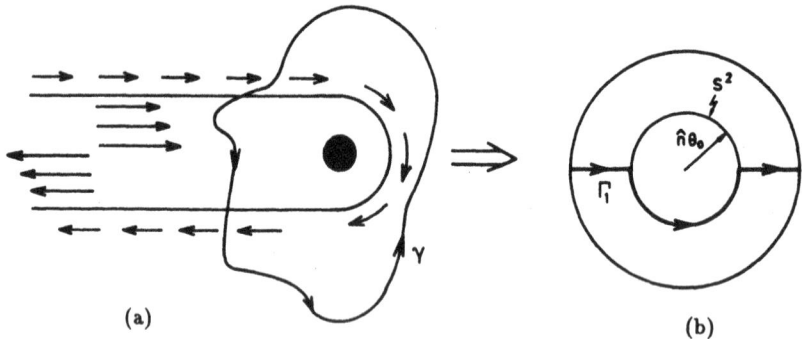

Fig. 7.14. (a) The $\hat{n}\theta$ field in the soliton terminating on the disclination. The closed contour γ around the disclination maps to the topologically nontrivial contour Γ_1 in (b) in the following way. Outside the soliton θ is fixed by spin-orbit coupling, $\theta = \theta_0$, and the $\hat{n}$ field performs the π winding. This corresponds to the motion on S^2. Within the soliton the vector $\hat{n}\theta$ closes the contour Γ_1 in the SO_3 space.

The situation is somewhat similar to the case of the Dirac monopole in the A-phase when the existence of the $\hat{l}$ hedgehog is incompatible with the topology of the *dreibein*. The solution of the paradox was that the hedgehog cannot exist as an isolated object but should be accompanied by the physical linear singularity (string). In the same manner the B-phase disclination line in the presence of the spin-orbit coupling cannot exist as an isolated object, it should be accompanied by the topological object of higher dimension, i.e. by some surface. The compromise between the small distance topology and the large distance topology of the disclination is thus achieved by the creation of the soliton attached to the disclination (see Fig. 7.14a). This soliton absorbs the extra pieces of the nontrivial contour Γ_1 in the SO_3 space outside the S^2 subspace in Fig. 7.14b. Everywhere outside the soliton wall of thickness of order dipole length ξ_d the angle $\theta = \theta_0$, while inside the wall θ changes along the pieces of the nontrivial contour in the SO_3 space.

Far from the disclination where the soliton structure depends only on one coordinate across the soliton, one may find this structure by solving the London equation

$$e_{\alpha\beta\gamma} R_{\beta i} \frac{\delta(F_{\text{grad}}^{\text{London}} + F_{\text{so}})}{\delta R_{\gamma i}} = 0 \ .$$

The simplest solution with $\hat{n} = \pm\hat{x}$ and with coordinate y across the soliton wall is as follows:

$$\tan \frac{1}{2}\theta = \sqrt{\frac{5}{3}} \coth \frac{y}{2\xi_d} \ . \tag{7.36}$$

This solution for the soliton may be unstable, but there is a special homotopy group which guarantees the topological stability of the soliton. So if the solution is unstable it will relax to a stable one within the same nontrivial topological class of solitons. This is the relative homotopy group $\pi_1(SO_3, S^2) = Z_2$, which contains two elements. They correspond to two nonequivalent classes of mapping of the line crossing the soliton body into the SO_3 space, with the ends of the line outside the wall, where the spin-orbital coupling reduces the freedom of the degeneracy parameters, being mapped into the S^2 subspace. The soliton belongs to the nontrivial class of this mapping, and is characterized by the same winding number $N_2 = 1$ as the corresponding disclination, at which the soliton can terminate. Due to this homotopy group the soliton can exist in the vessel regardless of the linear defect: it may extend from one wall of the vessel to another.

7.15. Topological Confinement of the Defects in Superfluid ^3He

In the general case, if the space R of the internal states is restricted by some internal or external perturbation to the subspace $\tilde{R}$, the topologically stable solitons arise if the relative homotopy group $\pi_1(R, \tilde{R})$ or $\pi_2(R, \tilde{R})$ is nontrivial. In Sec. 7.12 the restriction of the degeneracy states was produced by the container walls which gave rise to boojum described by the nontrivial elements of the group $\pi_2(R, R_{wall})$. So the hierarchy of interactions and length scales usually produces additional topologically stable textures. The corresponding calculation of $\pi_1(R_A, \tilde{R}_A)$ for the A-phase, where $\tilde{R}_A$ is the restricted manifold of internal states due to the spin-orbit coupling, shows that the $\hat{d}$ - $\hat{l}$ soliton, discussed in Sec. 3.7, is topologically stable and may terminate on the half-quantum vortices. So the half-quantum vortices in the A-phase are confined to each other by the solitonic film in the same manner as disclinations in the B-phase.

There are several examples of the topological confinement of defects in superfluid ³He. Four of them are illustrated in Fig. 7.15. In three of them the linear defects are confined by surfaces (films) terminating on defect, in the fourth example the point defects are connected by string (Fig. 7.15d). In the first three cases the attracting force due to the films does not depend on the distance between the defects, since the energy of the film is proportional to its length. In the case of the vortex string, which leads to attraction of the monopoles, the attractive force increases logarithmically in ³He-A and does not depend on the separation of monopoles in the case of a superconductor.

7.16. *Stabilization of the Vortex-disclination with the Solitonic Tail in the Rotating Vessel*

The combined object with $N_1 = 1$ ($N_1 = m_\Phi$) and $N_2 = 1$, which simultaneously has the properties of the quantized vortex, and that of the disclination with the solitonic tail, are stabilized in the rotating vessel due to this combination of the properties. On the one hand, as a vortex, it is forced by the Magnus force directed to the centre of the vessel to penetrate from the surface into the bulk rotating liquid

$$\vec{F}_{\text{Magnus}} = \rho_s \kappa N_1 \hat{z} \times (\vec{v}_s - \vec{v}_n) . \tag{7.37}$$

The Magnus force is absent in the equilibrium rotating state with the equilibrium density of conventional quantized vortices, which on average

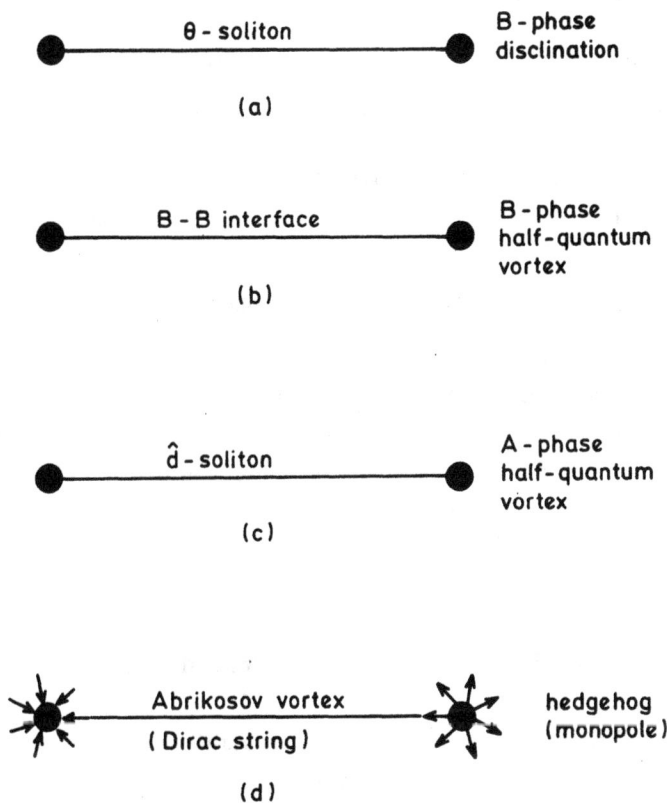

Fig. 7.15. Some examples of the topological confinement of the defects in superfluid ^{3}He. (a) Two disclinations in the B-phase are connected by the θ soliton. (b) Two half-quantum vortices comprising the nonaxisymmetric V2 vortex in the B-phase are connected by the B-B interface (see Sec. 8). (c) Two half-quantum vortices in the A-phase are coupled by the $\hat{d}$ soliton. (d) Two monopoles in the A-phase are confined by being connected by the vortex filament.

imitate the solid body rotation of the superfluid component in such a way that $< \vec{v}_s >= \vec{v}_n = \vec{\Omega} \times \vec{r}$, where $\vec{\Omega}$ is the angular velocity of the rotation of the vessel. So to observe the vortex-disclination one should create the metastable nonequilibrium vortex density. If the B-phase state is completely vortex free , then the conterflow $\vec{v}_s - \vec{v}_n = \vec{\Omega} \times \vec{r}$ produces the following dependence of the Magnus force, acting on the vortex disclination, on the distance r from the axis of the cylindrical vessel ($N_1 = 1$):

$$F_{\text{Magnus}} = -\rho_s \kappa \Omega r . \tag{7.38}$$

On the other hand the soliton, which, like the soap film, necessarily appears between the disclination line and the container wall, produces the surface tension in the opposite direction. This force tends to contract the film:

$$F_{\text{tension}} = E_{\text{soliton}} , \tag{7.39}$$

Here E_{soliton} is the energy of the soliton per unit area. If the angular velocity of the rotation, Ω, exceeds some critical value $\Omega_c = E_{\text{soliton}}/(\rho_s \kappa R)$, where R is the radius of the cylindrical vessel, then there is an equilibrium position r_0 of the vortex disclination in the vessel defined by the balance of the forces $\vec{F}_{\text{Magnus}} + \vec{F}_{\text{tension}} = 0$ (see Fig. 7.16):

$$\frac{r_0}{R} = \frac{\Omega_c}{\Omega} . \tag{7.40}$$

So if the vortex-disclination is somehow created, it occupies this position in the rotating vessel.

This object was created by special preparation of the nonequilibrium but highly stable rotating state. It was identified in NMR experiments which are sensitive to the size of the solitons, due to the specific dependence of the length $R - r_0$ of the soliton on the rotation velocity and on the size of the cluster of conventional vortex array (see Fig. 7.16).

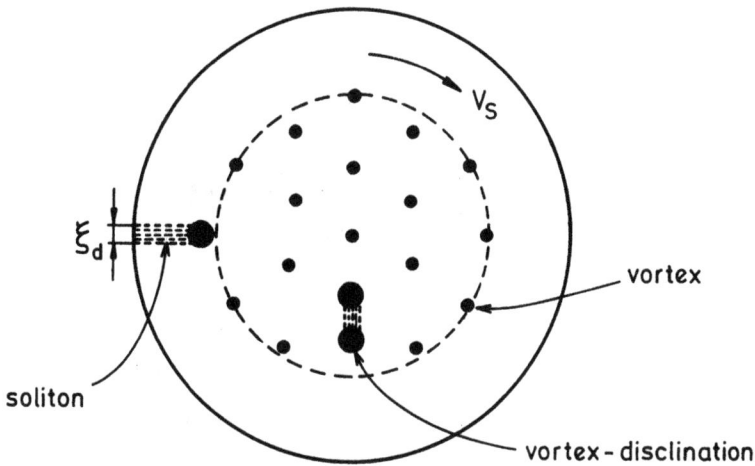

Fig. 7.16. The vortex-disclinations (large filled circles) obtained in the rotating vessel in a special way. Rotation was started in the A-phase, then under rotation the AB interface crosses the vessel leaving behind the rotating B-phase state with a cluster of conventional quantized vortices (small circles) and with a small fraction of the vortex-disclinations. Both the solitons connecting two vortex-disclinations and the solitons between the vortex-disclination and the wall are recorded. The appearance of these objects reflects complicated interaction of linear defects with the AB interface. This interaction is also regulated by the topology of the AB interface, which in particular gives rise to the monopole object in Fig. 7.11.

8
Spontaneous Symmetry Breaking in the ^{3}He Vortices

8.1. *Compulsory and Spontaneous Symmetry Breaking in Inhomogeneous Vacuum*

The topological object in condensed matter, as well as any other texture in the order parameter field, represents the inhomogeneous vacuum state, which is characterized by some residual symmetry. The texture violates some of the symmetries of the homogeneous vacuum state. The symmetry breaking may be dictated by the geometry of the object. For example translational symmetry is broken by the linear defect: only the translation along the line of the defect is the symmetry operation for this inhomogeneous vacuum, while other translations shift the line and therefore transform the state of the liquid with the linear defect into another degenerate state. This breaking of the translational symmetry leads to a new Goldstone mode which is absent in the homogeneous vacuum: this mode corresponds to elastic waves along the string, i.e. oscillations of the defect filament propagating along the line.

Another reason for the symmetry breaking is the incompatibility of the given symmetry with the given topological charge. Examples are: i) Violation of the time inversion symmetry by the quantized vortex in superfluids and superconductors. The time inversion operation changes the sign of the superfluid velocity circulation and therefore transforms the vortex with, say, $m_\Phi = 1$ into another degenerate state: a vortex with the opposite topological charge $m_\Phi = -1$. Such symmetry breaking within the vortex, combined with the spontaneously broken symmetry of the bulk liquid state, leads to new physical properties of the system, for example in the B-phase the quantized vortex acquires a spontaneous magnetic moment concentrated within the hard vortex core. This tiny magnetization of order 10^{-11} of the nuclear magneton per atom of rotating superfluid ^{3}He-B was experimentally resolved in NMR experiments.

ii) In the same manner the space parity P is broken in the continuous doubly quantized vortex in the A-phase, since P is incompatible with the requirement of the absence of singularity on the axis; this was discussed in Sec. 7.

iii) The breaking of the spherical symmetry of the A-phase monopole texture in the spherical vessel (see Secs. 3 and 7) is also the consequence of topological restrictions. This spherical symmetry is broken by the string terminating on the monopole.

So some symmetry is necessarily violated by the defect of a given geometry and topological class. However this is not the whole story. In addition, some symmetry breaking may occur spontaneously as in the homogeneous system. The maximally possible symmetry of the defect dictated by the topology suddenly breaks spontaneously if one changes some external parameter (magnetic field, pressure, temperature, the parameters β in the Ginzburg-Landau functional, etc.). Such situation takes place inside the hard core of the B-phase quantized vortex, where both the space parity and axial symmetry, which are good symmetries for the vortex, are nevertheless spontaneously broken. So one has two different types of vortices with the same topological charge $m_\Phi = 1$ but different residual symmetry and therefore with different physical properties. Like for the two kinds of condensed matter with different residual symmetries, there is phase transition between two vortex states, as observed in the NMR experiments on vortices. Since this phenomenon is extensively studied both theoretically and experimentally, we shall discuss the breaking of symmetry in the vortices in greater detail.

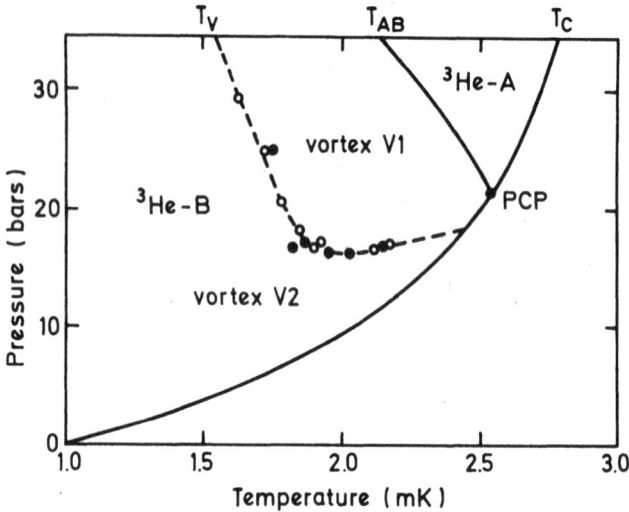

Fig. 8.1 The first order phase transition line $T_V(P)$ at which the vortex core transformation occurs. The line separates two species of quantized vortices, $V1$ and $V2$, which have the same winding number but different symmetries of the vortex core.

8.2. *Symmetry of Linear Defects*

A vortex solution may be rather complicated such that the minimization procedure involve trial functions with a large number of harmonics of the pairing amplitudes. Therefore one of the goals of the symmetry analysis is i) to reduce the number of trial functions to that which is relevant for a given symmetry class. Also, this analysis allows one ii) to describe the possible phase transitions, which should accompany the symmetry breaking, as well as iii) to characterize the distinct physical properties of the broken-symmetry states.

Symmetry classification of defects in condensed matter is somewhat analogous to the classification of the ordered states of the system themselves, though some difference exists. In order to find all the physically different defects of a given topological class one should enumerate all the possible symmetry groups H_{defect}, compatible with i) geometry of defect, with ii) its topological charge and with iii) the broken symmetry of the homogeneous ordered state far from the defect. All these constraints essentially restrict the number of possible symmetry classes of defects.

The first constraint shows that H_{defect} should be a subgroup of such a symmetry group G_{defect} of the physical laws in the presence of defect, which is relevant for the given geometry of defect. Here we consider linear defects. In the presence of the fixed line in the bulk liquid, the total symmetry group G_{line} of physical laws contains only such elements of the total group G which do not change the position of the line. The Euclidian group is now restricted to $D_{\infty h} \times t^z$, where $D_{\infty h}$ includes rotations around the axis of the linear defect $\hat{z}$, rotations by π around a perpendicular axis and space inversion P, and t^z denotes translations along $\hat{z}$. Thus the group G_{line} is

$$G_{\text{line}} = (D_{\infty h} \times t^z) \times (T \times U(1)) . \qquad (8.1)$$

The other two constraints are somewhat more complicated. Fortunately in practice one can avoid this formal scheme of obtaining the possible subgroups $H_{\text{line}} \in G_{\text{line}}$. Our simple scheme here is to find first the maximally symmetric subgroups H_{line} by considering the asymptotic behavior of the solutions of the Ginzburg-Landau equations. The total solution of the Ginzburg-Landau equations for the order parameter cannot be more symmetric than its asymptote far from the hard core region. And we know that outside the core the solution of the Ginzburg-Landau equations is the solution of the simpler London equations for the degeneracy parameter. Therefore our plan is i) to find the most symmetric solutions of the London equations for the degeneracy parameter outside the hard core; ii) to identify the symmetry groups of the obtained solutions, which are thus the most symmetric groups H_{line}; iii) to find the relevant harmonics of the order parameter, which are relevant for given maximal symmetry; and iv) to investigate possible breaking of maximal symmetry in order to describe the phase transitions in the defects.

8.3. Symmetry of Vortices in ^4He

We start with the vortices in the superfluid ^{4}He, whose maximal symmetry group proves to be applicable with some modification also to ^{3}He-B.

In the vortex with winding number m_Φ the solution of the London equation for the condensate phase Φ and for the order parameter ψ outside the hard core is

$$\psi(\vec{r}) = \text{const } e^{i\Phi(\vec{r})} = \text{const } e^{im_\Phi\phi} . \qquad (8.2)$$

This solution has two continuous and two discrete elements of symmetry. In addition to i) the translations t_z along the vortex axis another continuous symmetry is important. This is ii) the combination of the rotations about the axis $\hat{z}$ by angle α, $\phi \to \phi + \alpha$, accompanied by the gauge transformation $\Phi \to \Phi - m_\Phi\alpha$. For such physical quantities which are gauge-invariant, e.g. the superfluid velocity $\vec{v}_s(\vec{r})$ or the mass density ρ, this is simply the symmetry under rotation about the axis of the defect, i.e. axial symmetry (or axisymmetry). The generator of this combined transformation is

$$\mathbf{Q} = \mathbf{L}_z - m_\Phi\mathbf{I} . \qquad (8.3)$$

Superfluid He-II has no internal orbital structure, therefore the momentum projection operator contains only the spatial derivative:

$$\mathbf{L}_z = -i\frac{\partial}{\partial\phi} .$$

The other two symmetries are discrete: iii) The space parity P, combined with gauge transformation. The pure space parity transformation transforms $\phi \to \phi + \pi$, so the condensate function is multiplied by $\exp(i\pi m_\Phi)$, which can be compensated by the corresponding gauge transformation. For the gauge invariant variables this symmetry is a pure spatial parity. iv) The combined symmetry TU_2, which is the time inversion symmetry T combined with U_2, a rotation by π around the perpendicular axis. The pure symmetry T is not conserved in the vortex since the time-inversion operation changes the sign of the condensate phase Φ and therefore the direction of superflow circulation around the vortex axis ($Tm_\Phi = -m_\Phi$). Therefore, T may enter into a symmetry operation only in combination with other symmetry element, U_2, which changes the sign of the azimuthal angle ϕ in Eq.(8.2) and therefore restores the initial value of the order parameter ψ. One may check that these two discrete symmetries really leave invariant the superfluid velocity $\vec{v}_s(\vec{r}) \sim \hat{\phi}/\rho$ around the vortex core:

$$\mathbf{P}\left(\vec{v}_s(\vec{r})\right) = -\vec{v}_s(-\vec{r}) = \vec{v}_s(\vec{r}) \,, \quad \mathbf{T}\mathbf{U}_2\left(\vec{v}_s(\vec{r})\right) = -\mathbf{U}_2\vec{v}_s(\mathbf{U}_2\vec{r}) = \vec{v}_s(\vec{r}) \,.$$
(8.4)

Accordingly the product $\mathbf{P}\mathbf{T}\mathbf{U}_2$ of these symmetry elements is also the symmetry element of the vortex

$$\mathbf{P}\mathbf{T}\mathbf{U}_2\left(\vec{v}_s(\vec{r})\right) = \mathbf{U}_2\vec{v}_s(-\mathbf{U}_2\vec{r}) = \vec{v}_s(\vec{r}) \,.$$
(8.4a)

Let us find the general form of the order parameter which satisfies all the symmetries of the asymptotic solution of the London equation. The order parameter ψ, which is invariant under the continuous symmetries, should not depend on z and must satisfy the equation of the axial symmetry in Eq. (8.3) (remember that the elementary boson in the Bose condensate of superfluid ^{4}He is one atom of ^{4}He, therefore $\mathbf{I}\psi = \psi$):

$$\mathbf{Q}\psi \equiv \left(\frac{1}{i}\frac{\partial}{\partial\phi} - m_\Phi\right)\psi = 0 \,.$$
(8.5)

This equation has the following general solution

$$\psi(\vec{r}) = C(\rho)e^{im_\Phi\phi} \,,$$
(8.6)

where $C(\rho)$ denotes an arbitrary complex amplitude, which is a function of only the radial distance ρ from the axis of the line defect. This is a quantized vortex line with m_Φ quanta of circulation.

Since the bosons in He-II have no internal structure, all the angular momentum originates from the superflow around the vortex, that is why the total angular momentum quantum number L_z coincides with the circulation quantum number m_Φ for He-II quantized vortex. This quantum number is simultaneously the topological charge for the He-II vortex. The function $C(\rho)$ must vanish on the axis of the vortex in order to avoid a divergent kinetic energy; therefore, superfluidity is always broken on the axis of a He-II vortex, where all the vorticity is concentrated.

The most symmetric vortex should be invariant under the symmetries (8.4). The symmetry TU_2 requires that the function $C(\rho)$ in Eq. (8.6) for the He-II vortex should be real (apart from an arbitrary constant phase factor). The space parity P is automatically satisfied by this Ansatz.

8.4. The Most Symmetric Vortices in ³He-B

The vortices in ³He-B prove to be described by the same symmetry elements as in ⁴He with only one modification: the axisymmetry generator, the momentum projection, should be generalized to include internal orbital and spin rotations of the Cooper pair:

$$\mathbf{L}_z = -i\frac{\partial}{\partial\phi} + \mathbf{M}^L + \mathbf{M}^S . \qquad (8.7)$$

The axisymmetry equation (8.5)

$$\mathbf{Q}A_{\alpha i} \equiv (\mathbf{L}_z - m_\Phi)A_{\alpha i} = 0$$

may be easily solved if it is written in terms of the 9 complex (18 real) amplitudes $a_{MS,ML}$, i.e. $\mathbf{Q}a_{MS,ML} = 0$. This is because $a_{MS,ML}$ are amplitudes of the eigenstates for the Cooper pair spin and orbital momentum projections with the eigenvalues $M^S = -1, 0, +1$ and $M^L = -1, 0, +1$, and just these operators enter the total z projection in Eq. (8.7). The solution of the axisymmetry equation is

$$a_{MS,ML}(\vec{r}) = C_{MS,ML}(\rho)e^{i(m_\Phi - M^S - M^L)\phi} . \qquad (8.8)$$

Here $C_{MS,ML}$ are the functions of the distance ρ from the vortex axis. In the ³He-B vortices the nonzero $C_{MS,ML}(\infty)$ in the asymptotics are $C_{+-} = C_{00} = C_{-+}$ according to Eq. (2.4), and as follows from Eq. (8.8), each of them has the same phase factor $e^{im_\Phi\phi}$, which corresponds to the solution of the London equation for the asymptote:

$$A_{\alpha i}(\vec{r} \to \infty) = \Delta_B\delta_{\alpha i}e^{im_\Phi\phi} . \qquad (8.9)$$

Approaching the hard core these three amplitudes become independent and some other components also appear. As distinct from ⁴He, not all the C_{M^S,M^L} need to vanish at the origin: if any $M^S + M^L = m_\Phi$, then the corresponding $a_{M^S,M^L}(\vec{r})$ has zero winding number of the phase and therefore its amplitude $C_{M^S,M^L}(0)$ is not forced to vanish: $C_{M^S,M^L}(0) \neq 0$. This gives the possibility to have axisymmetric vortices which display no violation of superfluidity in their hard cores.

The discrete symmetries in Eq. (8.4) restrict the number of possible parameters $C_{M^S,M^L}(\rho)$ in the most symmetric vortex. The P symmetry operation transforms the functions $C_{M^S,M^L}(\rho)$ in the following way:

$$PC_{M^S,M^L} = (-1)^{M^S+M^L} C_{M^S,M^L} . \tag{8.10}$$

Thus the most symmetric vortex, which obeys $PC_{M^S,M^L} = C_{M^S,M^L}$, can have only even $M^S + M^L$, so only five functions C_{M^S,M^L} are nonzero, these are:

$$C_{++}, \quad C_{+-}, \quad C_{00}, \quad C_{-+}, \quad C_{--} . \tag{8.11}$$

Another restriction is imposed on the maximally symmetric ^{3}He-B vortices by the discrete symmetry TU_2 in Eq. (8.4). Through an appropriate choice of the common phase factor for the C_{M^S,M^L}, this transformation may be taken as:

$$TU_2 C_{M^S,M^L} = (-1)^{M^S+M^L} C^*_{M^S,M^L} , \tag{8.12}$$

while the PTU_2 symmetry operation is reduced to the complex conjugation:

$$PTU_2 C_{M^S,M^L} = C^*_{M^S,M^L} . \tag{8.12a}$$

Therefore, in the most symmetric vortices all the C_{M^S,M^L} are real (apart from the common constant phase factor). Thus the most symmetric vortices in superfluid ^{3}He-B are described by an integer m_Φ (the total angular-momentum quantum number, which coincides with the circulation quantum number), and have exactly five (from 18) independent real components $C_{M^S,M^L}(\rho)$ – in contrast with the single real function $C(\rho)$ for the vortices in He-II. The corresponding solution of the G-L equations for the maximally symmetric singly quantized vortex ($m_\Phi = 1$) is given in Fig. 8.2. All 5 components of the order parameter vanish on the vortex axis because

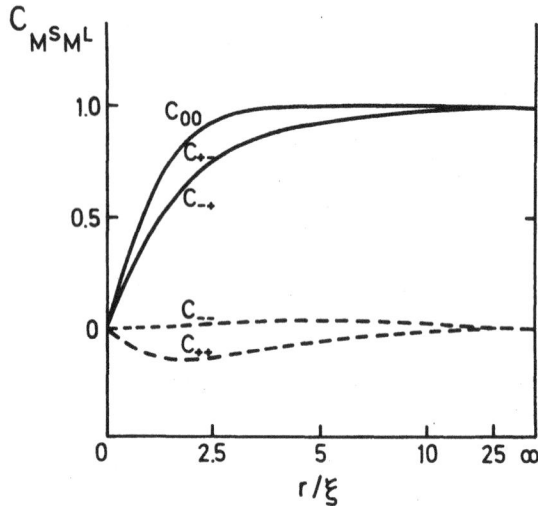

Fig. 8.2. The maximally symmetric singly quantized B-phase vortex has exactly five components of the order parameter within its hard core. All the components are zero on the vortex axis, which implies normal Fermi liquid on the axis. The component C_{++} corresponds to the A_1 phase in the core, which has ferromagnetically ordered spins along the vortex axis. This phase is responsible for the magnetic moment of the vortex core.

all of them have nonvanishing winding of the phase, $\exp[i(1 - M^S - M^L)\phi]$ (remember that $M^S + M^L$ is even).

It is important that even this most symmetric vortex possesses a magnetic moment of nuclear spins

$$M_z(\vec{r}) \sim \sum_{M^S, M^L} M^S \, |a_{M^S, M^L}|^2 = \sum_{M^L} (|\, C_{+,M^L}(\rho)\,|^2 - |\, C_{-,M^L}(\rho)\,|^2) \ .$$

$$(8.13)$$

The integral over the cross section of the vortex core gives a nonzero value, since there is no additional symmetry. The time-inversion T symmetry:

$$\mathbf{T}\, C_{M^S, M^L} = C^*_{-M^S, -M^L} \ , \qquad (8.14)$$

– which would make M_z equal to zero – is violated in the vortices. The spin rotation symmetry $SO_3^{(S)}$, which also prohibits the existence of spin magnetization, is already broken in the bulk liquid. So there is ferromagnetic ordering of the nuclear spins within the hard vortex core, induced by the time inversion symmetry breaking in the vortex.

This is a manifestation of the broken relative spin-orbit symmetry in the B-phase, discussed in Sec. 2. The vortex produces the orbital momentum of the liquid along $\hat{z}$ due to superflow circulation about the vortex axis. The broken relative spin-orbit symmetry implies that the appearance of the orbital momentum in the isotropic liquid should be accompanied by spin magnetizaton along $R_{\alpha i}\hat{z}_i$, where $R_{\alpha i}$ is the asymptote of the B-phase order parameter matrix far from the vortex. In our solution the asymptote was given by Eq. (8.9) and therefore the spontaneous nuclear magnetization was directed along the vortex axis. To obtain the solution with the arbitrary asymptote $\Delta_B R_{\alpha i} \exp(im_\Phi \phi)$ one should rotate the obtained solution in the spin space by matrix $R_{\alpha i}$. It is important that in this general case the vortex is magnetized in the direction $R_{\alpha i}\hat{z}_i$, defined both by the direction of the vortex axis and by orientation of the degeneracy parameter in the bulk liquid; this made it possible to observe this tiny magnetization of the vortex.

8.5. Broken Parity in ^{3}He-B Vortices

The maximally symmetric vortex is always a solution of the G-L equations and therefore corresponds to an extremum of the free-energy functional. However, it may appear that for the relevant β parameters in the G-L functional this is not a true minimum, but rather a saddle point. In ^{3}He-A the most symmetric doubly quantized vortex with singular core is unstable towards continuous vortex with broken space parity. In the same manner, the most symmetric vortex in ^{3}He-B, with all its pairing amplitudes being zero on the vortex axis and therefore with suppressed superfluidity in the core, may be unstable towards the restoration of superfluidity in the hard core, which is always accompanied by the breaking of space parity.

It is important that the symmetry P may be broken in two different ways, depending on which discrete symmetry in Eq. (8.4) is still retained, TU_2 symmetry or PTU_2. In the A-phase continuous vortex, described by Eq. (7.18), the combined symmetry PTU_2 is conserved. An example of the axisymmetric nonsingular A-phase vortex texture with the same quantization number $m_\Phi = -2$, but with conserved TU_2 symmetry is

$$\hat{l} = \hat{z}\cos\eta(\rho) \pm \hat{\phi}\sin\eta(\rho) \ . \tag{8.15}$$

This vortex has twofold degeneracy as well; the two states also transform into each other through space inversion P operation.

Moreover, it is possible that all the discrete symmetries in Eq. (8.4) are simultaneously broken due to energy considerations. The $\hat{l}$ texture in the corresponding axisymmetric vortex is given by

$$\hat{l} = \hat{z}\cos\ \eta(\rho) \pm \sin\ \eta(\rho)\big(\hat{\phi}\cos\alpha(\rho) \pm \hat{\rho}\sin\alpha(\rho)\big) \ .$$

This vortex has fourfold degeneracy. The four states are obtained from each other through the application of the elements P, TU_2 and PTU_2, which form the group $Z_2 \times Z_2$.

Estimation of the structure of these three A-phase vortices, TU_2-symmetric, PTU_2-symmetric and asymmetric, using simple trial functions shows that the TU_2-symmetric vortex is more preferable. This is not however the final result, since the complete numerical solution of the London equation for the A-phase vortices can change the situation. Also the result can depend on the temperature.

In the B-phase the exact numerical solutions of the G-L equations have been found for the singly quantized axisymmetric vortices near T_c. Computations show that in the whole pressure region, where the B-phase is stable, the minimum of the Ginzburg-Landau free-energy functional in the class of axisymmetric vortices with $m_\Phi = 1$ corresponds to the PTU_2-symmetric vortex with broken symmetry P. The most symmetric vortex proves to be unstable towards the formation of the vortex with broken parity and corresponds to a saddle-point of the Ginzburg-Landau functional.

In the PTU_2-symmetric vortex there is no constraint on $M^S + M^L$ and from Eq. (8.12a) it follows that all the components are real. So the vortex has nine real functions $C_{M^S,M^L}(\rho)$, see Fig. 8.3. Two of the amplitudes, $a_{0+} = \Delta_B C_{0+}$ and $a_{+0} = \Delta_B C_{+0}$, remain finite on the vortex axis. The largest of them corresponds to the A-phase, while the smaller one is the

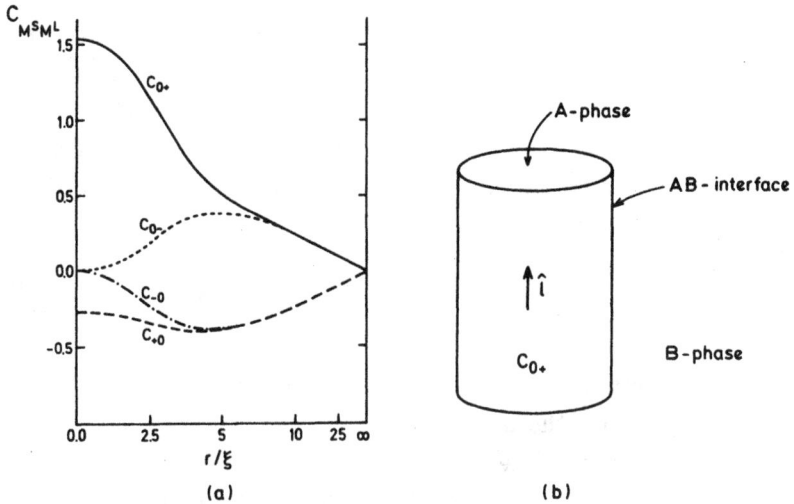

Fig. 8.3. (a) Four additional components in the singly quantized axisymmetric vortex with broken parity. This vortex corresponds to the $V1$ vortex in the region of the phase diagram in Fig. 8.1, which is close to the A-phase. The component C_{0+} is the most prominent in the core. This corresponds to the A-phase, which appears at the core due to the proximity effect and does not vanish on the vortex axis since it can support the continuous vorticity. So the core of the $V1$ vortex consists of the A-phase separated from the B-phase by the AB interface (b).

so-called β phase with nonzero spin projection on the vortex axis, $M^S = 1$. This β phase is principally responsible for the large magnetic moment of the PTU_2-symmetric vortex, as compared with that of the most symmetric vortex.

8.6. *A-phase Core of the ^{3}He-B Vortex*

Calculations show that close to the policritical point PCP, i.e., near the phase transition into the A-phase, the A-phase component in the hard core of the B-phase vortex nearly reaches its value in the pure bulk A-phase above transition. That is, the vortex core becomes nearly pure A phase, separated by a cylindrical layer of the AB interface from the B-phase in the bulk liquid (Fig. 8.3b).

So the hard core of the B-phase vortex may serve as the A-phase nucleation center. This means that in the presence of vortices, there should exist a catastrophe line in the (p, T)–plane at which the metastable B-phase becomes unstable towards the spontaneous growth of the A-phase core region. This may be illustrated by the following simple model. The vortex energy per unit length may be qualitatively expressed in terms of the vortex-core radius r_{core}:

$$E_{vortex} = \pi r_{core}^2 (F_A - F_B) + 2\pi r_{core} \sigma_{AB} + \pi \rho_s m_\Phi^2 \left(\frac{\hbar}{2m_3}\right)^2 \ln \frac{R}{r_{core}} . \quad (8.16)$$

The first term is the condensation energy of the A-phase core region, in comparison with the bulk B-phase energy, with $| F_A - F_B | \ll | F_B |$ near the AB-transition. The second term is the energy of the domain wall, with $\sigma_{AB} (\sim |F_B| \xi)$ denoting the surface energy of the interface between the A- and B-phases. The last term is the hydrodynamical energy of the mass superflow outside the vortex core, where R is the external cutoff radius, which does not enter the final result.

Minimization of E_{vortex} gives

$$r_{core} = \text{sign}(F_A - F_B) \sqrt{\left(\frac{\sigma_{AB}}{2(F_A - F_B)}\right)^2 + m_\Phi^2 \frac{\rho_s \left(\frac{\hbar}{2m_3}\right)^2}{2(F_A - F_B)}} - \frac{\sigma_{AB}}{2(F_A - F_B)} .$$

$$(8.17)$$

There are two consequences of this equation. i) In the region of the B-phase stability, i.e. when $(F_A - F_B) > 0$, the core size of the vortex with m_Φ circulation quanta increases with m_Φ. So in the vicinity of the transition one can expect the existence of the locally stable multiply quantized B-phase vortices. ii) In the metastable region, i.e. above the AB-transition where $(F_A - F_B) < 0$, the finite core size of the B-phase vortex exists only when the difference between the bulk B-phase energy and that of the A-phase does not exceed a critical value:

$$(F_B - F_A)_{cr} = \frac{\sigma_{AB}^2}{2\rho_s m_\Phi^2 \left(\frac{\hbar}{2m_3}\right)^2} . \quad (8.18)$$

At the critical threshold value (8.18), the vortex becomes unstable towards a spontaneous increase of the core radius to infinity, thus producing the transition to the A-phase without overcoming any energy barrier.

So in the vicinity of the AB-transition the most stable vortex is the broken parity axisymmetric vortex with $m_\Phi = 1$ and the A-phase core. From the above considerations it follows that there are two physical reasons why we have such a vortex: i) The ability of the A-phase to support continuous vorticity, due to this the singularity on the vortex axis can be dissolved; and ii) The proximity effect: it is more advantageous to have in the core the A-phase instead of the normal liquid, which exists in the most symmetric vortex. The A-phase has nearly the same energy density as the B-phase, while the normal component has essentially higher energy. The latter argument does not work far from AB-transition, as a result an additional symmetry breaking occurs at low pressure.

8.7. *Planar Phase vs. A-phase in the Core of the ^{3}He-B Vortex*

Numerical calculations show that at low pressures, far from the AB-transition, the A-phase core of the vortex proves to be unstable towards formation of a nonaxisymmetric core. The breaking of the axial symmetry of the vortex core is related to another tendency, which comes from the hidden symmetry of the BCS model. Incidentally the BCS model is a good approximation just at low pressures. Due to the hidden symmetry the metastable A-phase, which comprises the hard core of the B-phase vortex, has the same energy as another superfluid phase, the so-called planar state. And if the corrections to the BCS model are small, then $F_{planar} - F_A$ should also be small.

The planar state contains only two of the three components of the B-phase state in Eq.(2.4): $a_{+-} = a_{-+}$, while the third is exactly zero, $a_{00} = 0$. In terms of the conventional order parameter this is

$$A_{\alpha i}(\text{planar}) = \Delta_P(\hat{x}_\alpha \hat{x}_i + \hat{y}_\alpha \hat{y}_i) = \Delta_P(\delta_{\alpha i} - \hat{z}_\alpha \hat{z}_i) . \qquad (8.19)$$

The planar phase is described by some residual symmetry group H_{planar} which is not the subgroup of H_A or H_B. Therefore its order parameter is necessarily the solution of the homogeneous G-L equations. However the relevant β parameters are unfavourable for this phase, so this solution is a saddle point of the G-L functional in the homogeneous liquid. But nothing prohibits to have the planar state in the inhomogeneous situation of, say, the vortex core.

The fact that the planar and the A-phases have the same energy in the BCS model, can be seen from their fermionic spectra. The Bogoliubov matrix for the A-phase and the planar phase fermions follows are:

$$H(\text{A–phase}) = \tau_3(\epsilon(\vec{k}) - \mu) + \frac{\Delta_A}{k_F}\sigma_z(\tau_1 k_x - \tau_2 k_y) , \qquad (8.20a)$$

$$H(\text{planar}) = \tau_3(\epsilon(\vec{k}) - \mu) + \frac{\Delta_P}{k_F}\tau_1(\sigma_x k_x + \sigma_y k_y) . \qquad (8.20b)$$

The quasiparticle spectrum, which comes from the diagonal H^2, has the same structure for both liquids:

$$E_{\vec{k}}^2(A, P) = (\epsilon(\vec{k}) - \mu)^2 + \Delta_{A,P}^2 \frac{(k_x^2 + k_y^2)}{k_F^2} , \qquad (8.21)$$

and therefore the energies of the A and P phases, which are obtained by summation over all quasi particle energies in the fermionic vacuum, have the same dependence on the gap amplitude. Therefore in equilibrium $\Delta_P = \Delta_A$ and $F_{\text{planar}} = F_A$, in spite of the fact that these phases have different residual symmetries H of the group G of physical laws. What combines them in one multiplet is the hidden symmetry of the Bogoliubov matrix in the BCS approximation.

Now two phases with the same energy compete to represent the hard core of the B-phase vortex at low pressure: i) On the one hand the A-phase is favourable, since it supports continuous vorticity and therefore produces continuous unwinding of singularity on the vortex axis. But there is the domain boundary between B- and A-phases, which costs energy; ii) On the other hand there is no domain wall between the B-phase and the P-phase, since the latter can be obtained from the B-phase just by continuously decreasing the a_{00} component of the order parameter to zero. But the planar state cannot support continuous vorticity and unwinding of the singularity is impossible.

8.8. *Broken Axisymmetry in ^{3}He-B Vortex. Molecule of Half-quantum Vortices*

The final argument, which proves to make the planar state more preferable in the hard core, is the fact that like the A-phase, the planar state can support the half-quantum vortices. An example of the distribution of the P-phase order parameter around half-quantum vortex is:

$$A_{\alpha i}(x,y) = \Delta_P \left((\hat{x}_\alpha \cos m_d\phi + \hat{y}_\alpha \sin m_d\phi)\hat{x}_i \right.$$
$$\left. + (\hat{y}_\alpha \cos m_d\phi - \hat{x}_\alpha \sin m_d\phi)\hat{y}_i \right) e^{im_\Phi\phi} . \qquad (8.22)$$

As in the case of the A-phase half-vortex, if both the rotational winding number m_d and the vortex winding number m_Φ for the order parameter phase Φ are half-integers, then the order parameter field is continuous around the vortex axis.

So a model for the structure of the B-phase asymmetric vortex core at low pressure is the formation of the planar state in the hard core. Then the P-phase vortex splits into the pair of half-quantum vortices, thus breaking the axial symmetry (see Fig. 8.4). The vortices are confined in the molecule by the vortex sheet which consists of the planar state. The qualitative distribution of the order parameter outside the vortex sheet is given by

$$A_{\alpha i}(x,y) = \Delta_B \left((\hat{x}_\alpha \cos \frac{\phi_1 + \phi_2}{2} + \hat{y}_\alpha \sin \frac{\phi_1 + \phi_2}{2})\hat{x}_i \right.$$
$$\left. + (\hat{y}_\alpha \cos \frac{\phi_1 + \phi_2}{2} - \hat{x}_\alpha \sin \frac{\phi_1 + \phi_2}{2})\hat{y}_i + \hat{z}_\alpha\hat{z}_i \right) e^{i\frac{\phi_1+\phi_2}{2}} ,$$
$$(8.23)$$

where ϕ_1 and ϕ_2 are azimuthal angles around, respectively, the first and the second half-vortex in the molecule. While the planar components of the B-phase order parameter is continuous everywhere outside the half-vortices, the a_{00} component,

$$a_{00} = \Delta_B \exp i\frac{\phi_1 + \phi_2}{2} , \qquad (8.24)$$

has different signs on opposite sides of the sheet: $a_{00}(y = \pm 0) = \pm i\Delta_B$. Within the sheet the a_{00} component should be continuous, and in the simple model one has $a_{00}(y) = i\Delta_B \text{th}(y/\tilde{\xi})$, where $\tilde{\xi}$ is the thickness of the sheet.

So the vortex sheet represents some kind of soliton, the B-B wall, which separates two B-phase states with different orientation of the degeneracy parameters. Inside the B-B wall the a_{00} component crosses zero, which means that the B-phase transforms to the planar state in the middle plane of the soliton. In this simple model it was assumed that the thickness of the sheet is smaller than the intervortex distance. It can be shown that this model proves to be exact in the limit of a high magnetic field, where the B-phase is deformed and becomes close to the planar state.

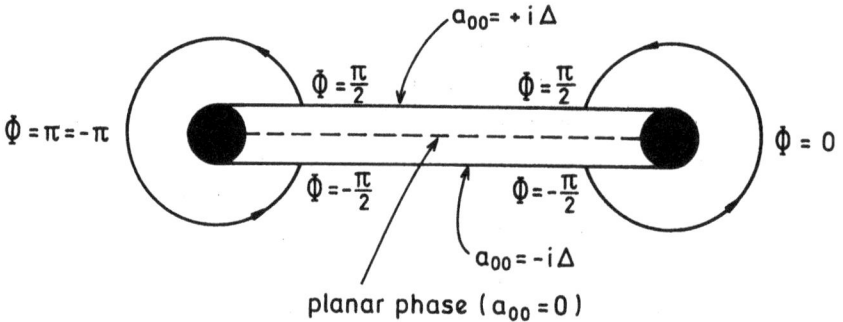

Fig. 8.4. The nonaxisymmetric $V2$ vortex as a molecule of two half-quantum vortices. The condensate phase Φ performs a π winding around each half-vortex. These vortices are connected by the vortex sheet – the domain wall between two bulk B-phases with different orientation of the degeneracy parameter, the B-B interface. One of the three components (a_{00}, a_{-+} and a_{+-}) comprising the B-phase, a_{00}, changes sign across the B-B interface and thus vanishes in the middle of the wall. This corresponds to the planar state within the vortex sheet.

8.9. Physical Properties of the Vortices with Broken Symmetry

Spontaneous breaking of symmetry in the soft core of the ³He-A and ³He-B vortices results in the appearance of new physical properties for these vortices, as compared with their counterparts in helium-4 and superconductors. In the vortices with broken space parity P, either a spontaneous electric polarization D_z arises along the vortex axis, or an axial supercurrent j_z. The situation depends on what is the residual symmetry of the vortex, TU_2 or PTU_2.

These two physical quantities have the following transformation properties under the symmetry operations P, TU_2 or PTU_2 in Eq. (8.4):

$$\mathbf{P}j_z = -j_z \,, \quad \mathbf{PTU_2}j_z = -j_z \,, \quad \mathbf{TU_2}j_z = j_z \,, \tag{8.25}$$

$$\mathbf{P}D_z = -D_z \,, \quad \mathbf{PTU_2}D_z = D_z \,, \quad \mathbf{TU_2}D_z = -D_z \,. \tag{8.26}$$

If there is P or PTU_2 symmetry in the vortex, then according to Eq. (8.25), the net supercurrent j_z through the cross section of the vortex-core region is exactly zero due to symmetry reasons. For example the P symmetry of the vortex requires that $j_z = \mathbf{P}j_z$ while the transformation property of the current is $\mathbf{P}j_z = -j_z$. These two requirements are compatible only if $j_z = 0$. Therefore, a nonzero spontaneous axial superflow in the core of the vortex may arise only if the P and PTU_2 symmetries are broken simultaneously. This is just the case for the vortex with residual TU_2 symmetry. The approximate solution of the London equations for the continuous A-phase vortices in the A-phase shows that just these TU_2 symmetric vortices seem to be more preferable than others. So the A-phase vortices should have spontaneous supercurrent along the vortex axis.

The B-phase vortices, both axisymmetric and asymmetric, are PTU_2 symmetric vortices. Therefore there is no spontaneous supercurrent, while the spontaneous electric polarization along the vortex axis is a specific characteristic of the B-phase vortices. The electric polarization in the vortices arises because of the so-called flexoelectric effect, which is well-known for ordinary (nonsuperfluid) liquid crystals. This polarization is caused by the bending of the order parameter, which produces a small deformation of the atomic ^{3}He shells.

The twofold degeneracy of the vortices with broken parity makes possible topologically stable kinks, or point solitons, separating the two parts of the vortex line with opposite supercurrent in TU_2 vortices in the A-phase and with opposite electric polarizations in PTU_2 vortices of the B-phase. Such a point on the B-phase vortex line has an electric charge, associated with it $e^* = D_z/2\pi$, which comprises a tiny fraction of the elementary charge of an electron. The breaking of axisymmetry in the core of the B-phase below the vortex core transition temperature, produces new properties related to the

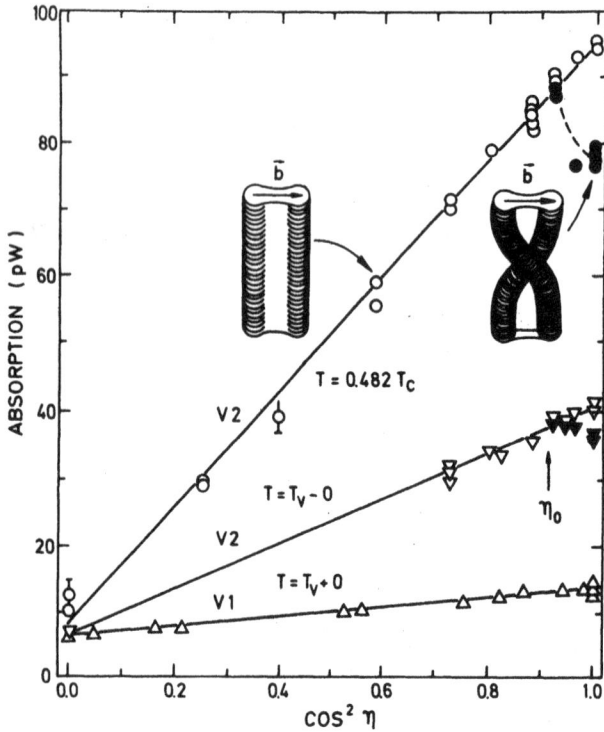

Fig. 8.5. NMR experiments which reveal the process of oscillations and twisting of the asymmetric vortices. Two branches are observed in the absorption of the nonlinear NMR mode on the nonaxisymmetric $V2$ vortices in ^{3}He-B as a function of the inclination angle η of the magnetic field towards the vortex axis. These branches correspond to untwisted and twisted states of the vortex. The precessing spins induce precession of the $\vec{b}$ vector, which, being pinned on the top and bottom walls of the container, makes the spiral state. The NMR absorption on twisted vortex state (filled circles at $T = 0.48T_c$ and filled inverse triangles for tempera-tures slightly smaller than the vortex transition temperature: $T = T_V - 0$) is less than that on an untwisted vortex state (open circles and open inverse triangles). The twisting occurs only at the small inclination angles: above some critical angle η_0 the $\vec{b}$ vector is completely pinned by magnetic field in untwisted state. No twisting is observed for the $V1$ axisymmetric vortex above T_V (empty triangles show absorption on $V1$ vortices for temperatures slightly larger than the vortex transition temperature: $T = T_V + 0$).

appearance of the preferred direction $\vec{b}$ in the vortex core, which shows the orientation of the molecule of half-quantum vortices. The breaking of continuous rotational symmetry of the vortex leads to an additional Goldstone mode for the vortex: oscillations of the $\vec{b}$ vector, which can propagate along the vortex axis. These oscillations as well as the spiral twisting of the $\vec{b}$ vector, induced by the precessing spins in the nonlinear regime of NMR, have been observed recently in NMR experiments on low pressure vortices, $V2$ vortices. This allowed one to identify the $V1$ and $V2$ vortices on the phase diagram (Fig. 8.1) as axisymmetric and nonaxisymmetric correspondingly (see Fig. 8.5).

9
Quasi-Two-Dimensional Superfluid ³He: Fractional Charge, Spin and Statistics

9.1. *Quasi-two-dimensional Objects in Superfluid* ^3He

Three-dimensional superfluid phases of liquid ^{3}He provide interacting quantum field systems with the maximal known broken symmetries in condensed matter physics. Consequently, a large number of different "elementary particles", fermions and bosons, can exist in these extraordinary condensed states, as well as stable nonuniform structures for the order-parameter field – *nonuniform vacua* – such as vortices, solitons, boojums, etc.... . These phases share many properties which were the sole privilege of particle physics, such as chiral anomaly, zero-charge effect, gauge fields, gravity, etc.

New concepts appear if one proceeds to lower dimensionality. For relativistic quantum field theory the two-dimensional system is not realistic and may serve only as a model, while for condensed matter physics this is reality with a number of new phenomena, such as integer and fractional quantum Hall effect (IQHE and FQHE), charge and spin fractionalization. In two-dimensional systems of superfluid ^{3}He with its unprecedented rich order parameter one may expect a wealth of unique phenomena which are impossible in other systems.

There are several different groups of quasi-two-dimensional systems in superfluid ^{3}He which have many features in common. These are:

1) The domain walls between two different bulk vacua of ^{3}He, such as i) A-B phase boundary; interfaces between the same superfluid states, but with different orientations of the degeneracy parameter: ii) B-B interface, which arises when one considers the structure of the asymmetric vortex in the B-phase, the B-B film couples the pair of half-quantum vortices in the vortex molecule (see previous section). In the same manner iii) the A-A interface can exist in the A-phase. For Fermi excitations which are localized on the wall the dynamics is purely two-dimensional and one may expect for them the anomalous behavior, which is characteristic for $(2 + 1)$-quantum field theory (QFT). This is a case of the B-B interface with a hard core of the order of the coherence length consisting of the planar phase. The fermionic excitations in the core of this interface have no energy gap providing an example of $(2+1)$ QFT with massless fermions.

In addition to the fermions the specific isolated point defects – boojums – can exist on the interface, which correspond to the particles with the topological charge in $(2 + 1)$-QFT. Such situation takes place for the A-B interface.

2) The superfluid states on the surface of the bulk superfluids. The more interesting situation arises here if the surface state is more degenerate than the bulk liquid. This may occur if for energetic reasons the anisotropic A-phase state takes place on the surface of the isotropic B-phase in the same manner as the A-phase appears in the core of the B-phase vortex. In this case in addition to the localized 2-dimensional fermions the isolated topological defects – the boojums – also can exist in the surface layer.

3) The superfluid films or superfluid surface states on the surface of the bulk normal liquid. They have the maximum number of degrees of freedom for the order parameter as compared with the other two-dimensional systems of ^{3}He since there is no coupling with the bulk order parameter. The superfluid states of ^{3}He, which are possible in thin films are the A-phase, the planar state and the A_1-phase. All of them have peculiar properties with fractional charge and/or fractional spin and statistics of the topological objects and with quantization of physical parameters leading to a variety of different types of QHE.

9.2. Superfluid Phases in Thin Films

Here we consider very thin films of superfluid ³He on a substrate. In order to observe the variety of the effects of the parameter quantization resulting from the quasi two-dimensional nature of the film, its thickness should be at least less than 20–30 interatomic spaces or 100 Å, which is less than the coherence length for superfluidity. To preserve the p-wave superfluidity in such conditions the requirement of perfect smoothness of the substrate surface should be fulfilled.

As in the bulk liquid, in the ³He film also two superfluid phases compete in equilibrium. The existence and symmetry of the superfluid phases in films are restricted by the boundary conditions, which require that the transverse components of the order parameter vanish on the boundary: $A_{\alpha z} = 0$, where z is the axis along the normal to the film.

The phases which survive this condition and have low energy are ³He-A and the planar state, discussed in Sec. 7. The A-phase state survives, since in order to meet the boundary conditions it is enough to orient the $\hat{l}$ vector along the normal. For the B-phase the boundary conditions require that the a_{00} component of the order parameter in Eq. (2.4) should vanish, which means transformation to the planar state.

9.3. Generations of Fermions

In the film the transverse motion of the quasiparticles – along the normal to the film – is quantized, as a result the Bogoliubov matrix acquires also the indices nn' of the transverse levels

$$\hat{H}_{nn'}(\vec{k}) = \begin{pmatrix} \varepsilon_{nn'}(\vec{k}) & \Delta_{nn'}(\vec{k}) \\ \Delta^{\dagger}_{nn'}(\vec{k}) & -\varepsilon_{nn'}(\vec{k}) \end{pmatrix} . \tag{9.1a}$$

Here $\vec{k}$ is a two-dimensional vector and $\varepsilon_{nn'}(\vec{k})$ is the matrix, in which the off-diagonal elements describe the interaction between the particles from different levels in normal Fermi liquid.

In the simplest case when the off-diagonal interaction is absent one has

$$\varepsilon_{nn'}(\vec{k}) = (\varepsilon_n(\vec{k}) - \mu)\delta_{nn'}$$

where in Fermi gas approximation

$$\varepsilon_n(\vec{k}) = \frac{\pi^2 n^2 \hbar^2}{2m_3 a^2} + \frac{k^2}{2m_3} . \qquad (9.1b)$$

Here $\varepsilon_n(0)$ is the nth energy level of transverse motion and a is the film thickness. In the case of normal liquid above T_c the fermions filling the negative levels, $\varepsilon_n(\vec{k}) - \mu < 0$, form several two-dimensional Fermi-liquids. In the simple case of the noninteracting levels the number n_0 of the independent Fermi systems is the largest n, at which $\varepsilon_n(0) < \mu$. This is the number of generations of fermions in analogy with particle physics, which arise due to compactification of the third dimension. Each two-dimensional Fermi liquid has its own Fermi momentum k_{Fn}, which in the simple model is

$$k_{Fn}^2/2m_3 = \mu - \varepsilon_n(0) . \qquad (9.1c)$$

Below T_c each generation of fermions acquires its own order parameter, which for the A-phase state has the form of Eq. (5.11) with vectors $\hat{e}^{(1)}$ and $\hat{e}^{(2)}$ being parallel to the plane of the film:

$$\Delta_{nn'}(\vec{k}) = \delta_{nn'} \frac{\Delta_n}{k_{Fn}} \vec{\sigma} \cdot \hat{d} (k_x \pm i k_y) e^{i\Phi} . \qquad (9.1d)$$

Due to interactions between the generations of fermions (fermions from different levels), they have a common orbital vector $\vec{l}$ fixed along the normal to the film, $\hat{l} = \pm \hat{z}$; a common condensate phase Φ which is well defined since the $\hat{l}$ vector is fixed; and a common $\hat{d}$ vector of spin antiferromagnetism, which is free to rotate in the film, since the spin-orbit coupling is very small.

The most important difference between the bulk A-phase and the A-phase film is that the quasiparticle spectrum in the film has no more gap nodes due to the quantization of the motion along $\hat{l}$. In the bulk liquid the Fermi point takes place for $\vec{k} \parallel \hat{l}$. Since the momentum $\vec{k}$ is two-dimensional in the film, it cannot be oriented along $\hat{l}$ to obtain a Fermi point. The quasiparticle energy on the nth level in the simplest model of noninteracting levels,

$$E_n^2(\vec{k}) = (\varepsilon_n(\vec{k}) - \mu)^2 + \left(\frac{\Delta_n}{k_{Fn}}\right)^2 k^2 , \qquad (9.2)$$

is nowhere zero. Only when one changes the external parameters and one of the levels, say n_1, of the transverse motion crosses the Fermi level $\varepsilon_{n_1}(0) - \mu$, then at the moment of crossing the spectrum $E_{n_1}(\vec{k})$ vanishes at $k = 0$. Therefore the dynamics of the A-phase in the film has no singularities and anomalies of bulk A-phase dynamics. Nevertheless the A-phase film still has nontrivial topology in momentum space, which is the reminiscence of the topologically stable Fermi points in the bulk A-phase. This nontrivial topology results in many exotic physical properties of the films.

9.4. *Symmetry and Internal Topology of Ground State*

Here we encounter an important concept of the momentum space topology. This gives, in addition to the symmetry classification of the different states of condensed matter, a finer classification. The phases may have the same residual symmetry H, but nevertheless there is still some important differences in their properties.

Let us consider two examples. i) Two crystals belonging to the same crystal symmetry group, but nevertheless one of them is a metal, while the other is an insulator. The difference comes from the different properties of the quasiparticle spectrum: the former has a Fermi surface, while the latter does not. This proves to be the difference in the internal topology of the two systems, since the Fermi surface is some topological singularity in the momentum space for the Green's function which describes the electronic properties of the system.

What is important here is that the difference between the sytems with the same symmetry but with different topologies of the electronic states, manifests itself only in the limit of zero temperature: in both cases the conductivity is non-zero, but has different asymptotic behavior when $T \to 0$. So the phase transition between these two systems is the Lifshitz transition which takes place at $T = 0$ when some other external parameter (pressure or magnetic field) changes (see Fig. 9.1).

ii) Another example was considered in Sec. 6.2, where the Lifshitz zero-temperature transition occurs between the anomaly-free state of the A-phase to the A-phase state with Fermi points. The transition occurs when the chemical potential crosses zero value. This is analogous in some sense to the

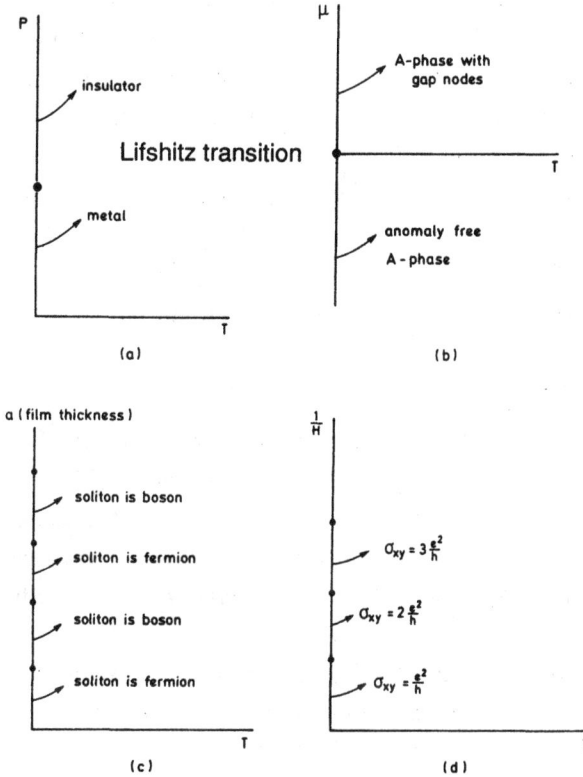

Fig. 9.1. Examples of the zero temperature Lifshitz transition, which is not accompanied by the symmetry breaking but manifests the change in internal topology. a) Metal-insulator transition, at which the Fermi surface disappears. b) Transition which occurs in the A-phase state if the chemical potential μ crosses zero. The anomaly-free state without gap nodes exists at $\mu < 0$, at $\mu > 0$ two topologically stable Fermi points appear. c) The change in the quantum statistics of the solitons in ^{3}He-A films occurs if the thickness a of the film crosses some critical values (see below). d) The quantum number for the Hall conductivity in QHE changes at some critical values of magnetic field H.

metal-insulator transition: the nondissipative state with a gap transforms to the dissipative state with two Fermi-points in momentum space.

In the film we have a similar situation: while the symmetry of the bulk A-phase state and the symmetry of the A-phase in thin film are the same (if one fixes the orientation of the $\hat{l}$ vector in bulk liquid), the Fermi points are absent in the film due to the dimensional quantization of the transverse levels. So when one decreases the thickness of the film from infinity to some finite value one may expect a sequence of Lifshitz transitions, when the transverse levels cross the chemical potential.

In all these cases the fermionic spectrum has distinctive topological properties below and above Lifshitz transition, and we may now quantitatively describe the difference in the topology, using Green's function describing the properties of the fermionic degrees of freedom.

9.5. *Topological Invariant for the Fermi Point*

Let us begin with the topological properties of the Fermi points in the momentum space of the bulk A-phase. The topology of the Fermi point in the 3-dimensional momentum space is similar to the topology of point singularities in 3-dimensional coordinate space. As we have seen from Sec. 7.10 the singular points in the real space, hedgehogs, are described by the integer topological invariant, which identifies the class of mapping of the surface, embracing the defect, onto the degeneracy parameter space. For the simplest case of the degeneracy parameter, the unit vector $\hat{d}$, the topological charge is given by the analytical expression (7.22).

A similar expression may be applied to Fermi points in the A-phase. First note that Eq. (7.22) may be generalized to treat not only the field of the unit vector $\hat{d}$ but also any vector field $\vec{d}(\vec{r})$ with only one condition: $\vec{d}(\vec{r}) \neq 0$. For the nonzero $\vec{d}$ one can construct the unit vector $\hat{d} = \vec{d}/|\vec{d}|$, and substituting this into Eq. (7.22) one obtains the topological invariant for the nonzero $\vec{d}$ field in the form

$$\tilde{m}_d = \frac{1}{8\pi} \int_{\text{around } \vec{r}_0} dS^i e_{ikl} \mid \vec{d}(\vec{r}) \mid^{-3} \left(\vec{d} \cdot \frac{\partial \vec{d}}{\partial x_k} \times \frac{\partial \vec{d}}{\partial x_l} \right) . \quad (9.3)$$

Here $\vec{r}_0$ is the position of the point singularity. At this singular point $\vec{r}_0$ the unit vector $\hat{d}$ is not defined which means that $\vec{d}$ vanishes, $\vec{d}(\vec{r}_0) = 0$. So the integer topological invariant Eq. (9.3) describes the topologically stable zeroes of the vector field $\vec{d}(\vec{r})$.

Let us apply this to zeroes in the quasiparticle spectrum in momentum space. Let us begin with 2×2 Bogoliubov matrix, which in the A-phase describes the fermions with given spin projection $M_S = (1/2)\vec{\sigma} \cdot \hat{d}$ (see Eq. (5.17)). The general form of the 2×2 Bogoliubov matrix may be described in terms of the vector function $\vec{m}(\vec{k})$ in momentum space:

$$\mathbf{H} = \vec{\tau} \cdot \vec{m}(\vec{k}) . \tag{9.4}$$

The A-phase matrix in Eq. (5.17) is a particular case of the function $\vec{m}(\vec{k})$ with components

$$m_3(\vec{k}) = \epsilon(\vec{k}) - \mu , \quad m_1(\vec{k}) = 2M_S \frac{\Delta_A}{k_F} \hat{e}^{(1)} \cdot \vec{k} , \quad m_2(\vec{k}) = -2M_S \frac{\Delta_A}{k_F} \hat{e}^{(2)} \cdot \vec{k} . \tag{9.5}$$

The zeroes of this vector function $\vec{m}(\vec{k})$ correspond to the zeroes in the fermionic spectrum since $E^2(\vec{k}) = \mathbf{H}^2 = \vec{m}^2(\vec{k})$.

So to describe the topologically stable Fermi point, point $\vec{k}_0$ where the zero of the fermionic spectrum occurs, $E(\vec{k}_0) = 0$, one may apply the expression (9.3) to the vector function $\vec{m}(\vec{k})$ in momentum space:

$$\tilde{m} = \frac{1}{8\pi} \int_{\text{around } \vec{k}_0} dS^i e_{ikl} \mid \vec{m}(\vec{k}) \mid^{-3} \left(\vec{m} \cdot \frac{\partial \vec{m}}{\partial k_k} \times \frac{\partial \vec{m}}{\partial k_l} \right) . \tag{9.6}$$

Now it is possible to check that the point zero $\vec{k}_0 = k_F \hat{l}$ in the A-phase is really topologically stable: the calculation of its topological charge in Eq. (9.6), using Eq. (9.5), gives nonzero charge $\tilde{m} = -1$ for each of two spin projections M_S. The same charge corresponds to the zero $\vec{k}_0 = 0$ in the spectrum of the left chiral fermions with

$$\mathbf{H} = -c\vec{\tau} \cdot \vec{k} ,$$

where $\vec{m}(\vec{k}) = -c\vec{k}$. Therefore these two systems, the bulk A-phase with Fermi points and quantum electrodynamics with massless chiral fermions, have a lot in common, including the same phenomenon of chiral anomaly.

For the node on the opposite pole with $\vec{k}_0 = -k_F\hat{l}$, the topological charge is opposite, $\tilde{m} = 1$, which is the same as the topological charge of zero $\vec{k}_0 = 0$ in the spectrum of the right chiral fermions.

9.6. *Topological Invariant in Terms of the Green's Function*

In the general case the Bogoliubov Hamiltonian is no more the 2×2 matrix and therefore cannot be expressed in terms of the vector field $\vec{m}(\vec{k})$. In superfluid ^3He this is a 4×4 matrix. In crystals the index of the energy band should be also added, in the ^3He films the Hamiltonian matrix is also described by the index n of the transverse levels. Moreover, if the interaction between the quasiparticles is important the single particle Hamiltonian is no longer relevant for the description of the fermionic excitations. So the expression (9.6) for the topological charge should be generalized to include all these physical complications.

The relevant physical quantity, which instead of the one-particle Hamiltonian describes the properties of the interacting electrons in crystals and fermionic quasiparticles in ^3He, is the Green's function $\mathbf{G}(i\omega, \vec{k}) = G_{nn'}(i\omega, \vec{k})$, where n enumerates all the possible indices (spin, band, etc.), and ω is the imaginary frequency. For the simplest case of the noninteracting fermions:

$$\mathbf{G}(i\omega, \vec{k}) = \frac{1}{i\omega + \mathbf{H}(\vec{k})} \ . \tag{9.7}$$

The Green's function is a function in the four-dimensional momentum space $k^\mu = (\vec{k}, \omega)$. Even from the simplest equation (9.7) one may conclude what is the most important region in this 4-dimensional space. The Fermi-manifold, which gives all the low-temperature behavior of the fermions, is defined as a set of points $\vec{k}_0$ where the Hamiltonian has zero eigenvalues, $E(\vec{k}_0) = 0$. In the Green's function these zeroes can be manifested only when $\omega = 0$, but for these points $k_0^\mu = (\vec{k}_0, 0)$ in four-dimensional space the determinant of the Green's function matrix becomes infinite, $\det \mathbf{G}^{-1}(\vec{k}, \omega) = 0$. At these points the Green's function has no inverse matrix. So this manifold represents the singular hyper-surface for the Green's function in the 4-dimensional momentum space, and in the general case, when Eq. (9.7) does not hold, the equation $\det \mathbf{G}^{-1}(\vec{k}, \omega) = 0$ defines the singular hyper-surface in the momentum space which is responsible for the low-temperature behavior of the Fermi system.

The singular two-dimensional surface in $(\vec{k}, \omega)$ space in this description just represents the conventional Fermi surface in the $\vec{k}$ space. Since the two-dimensional singularities of the Green's function in the four-dimensional space prove to be topologically stable (they are described by the same homotopy group π_1 as the vortices in real space), this explains why the Fermi surface is such a stable formation. The topological stability of the Fermi surface is in the basis of the Landau theory of the Fermi liquid: Fermi surface survives if the interaction between the particles is adiabatically switched on. Here we are interested in the Fermi points, in general description in terms of the Green's function they correspond to the isolated singular points in the 4-dimensional $(\omega, \vec{k})$ space, which are described by another homotopy group.

The topological charge (9.6) in terms of the vector field, which describes the isolated Fermi points in the simplest case of 2×2 Hamiltonian of noninteracting fermions, in the general case may be written in terms of the Green's function matrix in the following way:

$$\tilde{m} = \frac{1}{24\pi^2} e_{\mu\nu\lambda\gamma} \text{tr} \int_{\text{around } \omega=0, \vec{k}=\vec{k}_0} dS^\gamma \mathbf{G} \partial_{k^\mu} \mathbf{G}^{-1} \mathbf{G} \partial_{k^\nu} \mathbf{G}^{-1} \mathbf{G} \partial_{k^\lambda} \mathbf{G}^{-1} .$$
$$(9.8)$$

Here $\mu = 0, 1, 2, 3$ and $k^0 = \omega$, $k^1 = k_x$, $k^2 = k_y$, and $k^3 = k_z$.

Equation (9.8) is a true topological invariant since it describes the mapping of the three-dimensional sphere S^3 around a singular point in the four-dimensional k^μ space onto the space R of the nonsingular (nondegenerate) matrices. This corresponds to the homotopy group π_3, which in the case of the nonsingular matrices is the group of integers, $\pi_3(R) = Z$. One may check that in the simple case of Eq. (9.7) of noninteracting fermions, with the 2×2 Hamiltonian in Eq. (9.4), Eq. (9.8) is reduced to the topological invariant Eq. (9.6) for the Fermi point $\vec{k}_0$ in terms of the vector $\vec{m}(\vec{k})$. So Eq. (9.8) gives integer topological quantization for the Fermi points in an arbitrary Fermi system (normal metals, superconductors, superfluids, antiferromagnets, etc.).

9.7. *Topology of the Ground State of the ^{3}He Film*

Now we can return to the ^{3}He-A film. The topology of the ground state of the film is related to the topology of the Fermi point in the bulk liquid. Let us consider first the case of the noninteracting levels. The integral over the spherical surface around $\vec{k}_0 = k_F \hat{l}$ in Eq. (9.6) may be represented as the difference of the integrals over the planes $k_z = $ const below and above the Fermi point

$$\int_{\text{around } k_F \hat{l}} = \int_{k_z > k_F} dk_x dk_y - \int_{-k_F < k_z < k_F} dk_x dk_y . \qquad (9.9)$$

Since the unit vector $\hat{m} = \vec{m}(\vec{k})/ \mid \vec{m}(\vec{k}) \mid$ is fixed at $\vec{k} \to \infty$, i.e. $\hat{m}(\infty) = (0,0,1)$ (see Eq. (9.5)), each of the two integrals on the right-hand side of Eq. (9.9) is also an integer topological invariant, which does not depend on the position of the plane, until the plane crosses the Fermi point. The first integral is exactly zero, since the plane can be continuously shifted to $k_z = +\infty$, where the unit vector $\hat{m}$ is constant and therefore does not contribute to the integral. So the second integral over any plane in the interval $-k_F < k_z < k_F$ equals 1.

The integral over the (k_x, k_y) plane is just the characteristic of the two-dimensional fermions on the filled transverse level in the film. So each of the n_0 transverse levels below the chemical potential level, $\varepsilon_n(0) < \mu$, has the same nontrivial topological invariant, which in terms of the 4×4 Green's function $\mathbf{G}_n$ for this level is:

$$N_n = \frac{1}{24\pi^2} e_{\mu\nu\lambda} \mathrm{tr} \int dk_x \, dk_y \, d\omega \, \mathbf{G}_n \partial_{k^\mu} \mathbf{G}_n^{-1} \mathbf{G}_n \partial_{k^\nu} \mathbf{G}_n^{-1} \mathbf{G}_n \partial_{k^\lambda} \mathbf{G}_n^{-1}$$

$$= 2(\vec{l} \cdot \hat{z}) ,$$

$$(9.10)$$

where $\nu = 0, 1, 2$. The value of this integral depends on the orientation of the $\hat{l}$ vector: $\hat{l} = \pm \hat{z}$; the factor 2 arises if one takes into account that each level has the 2-fold spin degeneracy. Note that in this two-dimensional system there is no Fermi manifold, the quasiparticle energy is nowhere zero. Nevertheless the ground state of the system has nontrivial internal topology.

Now let us adiabatically switch on the interaction between the levels. The individual charge N_n loses its meaning, while the total charge of the ground state of the film remains to be well defined. It has the following form in terms of the general Green's function matrix $\mathbf{G} = G_{nm}$ with all the indices:

$$N = \frac{1}{24\pi^2} e_{\mu\nu\lambda} \text{tr} \int dk_x \, dk_y \, d\omega \; \mathbf{G}\partial_{k_\mu}\mathbf{G}^{-1}\mathbf{G}\partial_{k_\nu}\mathbf{G}^{-1}\mathbf{G}\partial_{k_\lambda}\mathbf{G}^{-1} \, . \quad (9.11)$$

It transforms to $2n_0(\vec{l} \cdot \hat{z})$ in the limit of n_0 noninteracting levels, i.e. to the sum $\sum_n N_n$ over the filled Landau levels.

9.8. Adiabatical Process, Conservation of Topological Invariant and Diabolical Points

The topological charge N in Eq. (9.11) does not change under any adiabatical perturbations of the order parameter or of external conditions, including adiabatic switching off the interlevel interaction. So $|N|/2$ is the number of the transverse levels below μ in the noninteracting Fermi-gas state, from which the real Fermi-liquid state is obtained when the interaction is adiabatically switched on. The adiabatical process here means that in this process the system passes only the nondissipative states (states with a gap in the quasiparticle spectrum), so the quasiparticle energy spectrum never touches the chemical potential level during the adiabatical process.

As an example of the nonadiabatic process one can consider the case when one of the levels of the transverse motion crosses the chemical potential, $\varepsilon_{n_1}(0) = \mu$ (Fig. 9.2). At the moment of crossing the quasiparticle energy spectrum touches zero at the point $k_x = k_y = 0$, i.e. the system passes through the intermediate dissipative state, which violates the adiabaticity condition. At that moment the topological invariant abruptly changes from N to $N \pm 2$ (Fig. 9.3). This bifurcation corresponds to the Lifshitz transition.

This reminds one of the phenomenon of integer Quantum Hall Effect, when the plateau of the integer value of the Hall conductivity $\sigma_{xy} = N\frac{e^2}{h}$ takes place only when there is no dissipation, i.e. the longitudinal conductivity $\sigma_{xx} = 0$. The transition between states with different adiabatical invariants N occurs through the dissipative state with $\sigma_{xx} \neq 0$.

On the other hand this is a familiar process of the levels crossing. Let us introduce some external parameter u, which we can continuously change.

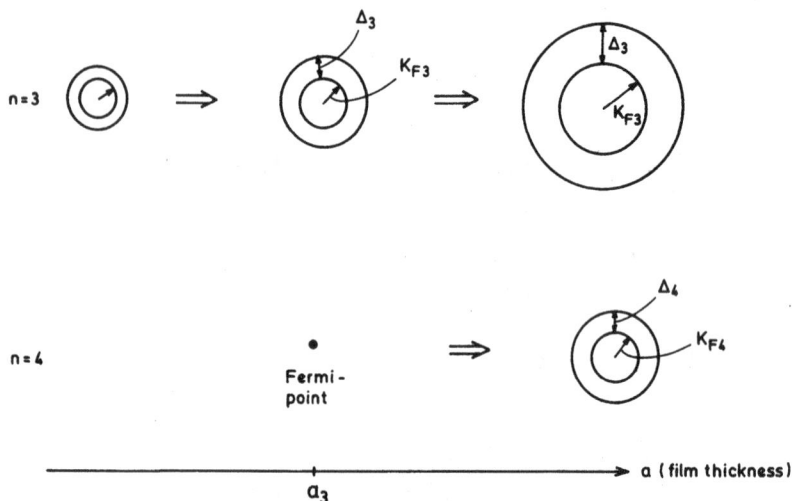

Fig. 9.2. Creation of a new generation of fermions in the film. Two levels of transverse motion, with $n = 3$ and $n = 4$, are shown as a function of the film thickness a. The fermionic vacuum on the fourth level is created at the critical thickness $a = a_3$ when the energy of the fourth level crosses the chemical potential. At $a = a_3$ the spectrum of the fermions on the fourth level has the Fermi point (diabolical point of the quasiparticle spectrum). At $a > a_3$ a conventional quasiparticle spectrum is developed on the fourth level with its own Fermi-momentum k_{F4} and a gap Δ_4.

This may be the chemical potential μ or the thickness a of the film. Together with the components k_x and k_y of the 2-dimensional momentum $\vec{k}$ this gives three parameters, on which the quasiparticle spectrum depends: $E = E(u, k_x, k_y)$. It is known from the non-crossing theorem, that the branches of the spectrum generally cannot cross each other, but in the 3-dimensional space of parameters there can exist isolated point (conical or diabolical point) where two branches can touch each other. This point is topologically stable and is described by the invariant (9.6) in 3-dimensional space (u, k_x, k_y), where the Hamiltonian H describes the interaction of two branches. In our case this Hamiltonian describes the interaction of the particles and holes. So the branches, which touch each other in our case, are the particle branch $(E > 0)$ and the hole branch $(E < 0)$. When they touch each other at point (u_0, k_{x0}, k_{y0}) their energy E becomes zero at this point,

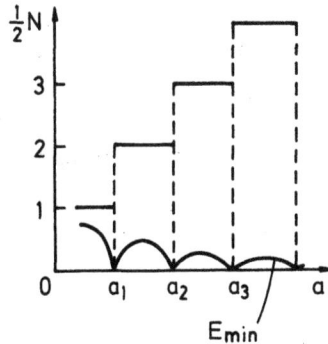

Fig. 9.3. Abrupt change of the internal topological invariant N occurs at critical values of the film thickness a, where the system crosses the intermediate dissipative states with the diabolical points in the quasiparticle spectrum. E_{min} denotes the minimal value of the quasiparticle energy which is zero at critical values of a.

which just corresponds to the intermediate dissipative state.

So the nondissipative ground state of the 2-dimensional system is characterized by one or several topological charges, which do not change under external adiabatical perturbations. The jump between the states with the different values of topological charge occurs by passing through the intermediate dissipative state. Such transition occurs only at zero temperature, being some kind of Lifshitz transition.

9.9. *Quantum Statistics of Solitons and θ-term in Action*

There are two groups of phenomena resulting from the existence of the integer topological invariant N in momentum space. The first group is related to the properties of topological particle-like solitons in the field of the $\hat{d}$ vector. These solitons are characterized by an integer-valued topological invariant $\tilde{m}_d$ in the real space x, y of the film (see Eq. (7.21)). The simplest realization of the soliton, also known as skyrmion, with the topological charge $\tilde{m}_d = 1$ is the axially symmetric fountain-like $\hat{d}$-texture

$$\hat{d}(x, y) = \hat{z}\cos\beta(\rho) + \hat{\rho}\sin\beta(\rho) , \qquad (9.12)$$

with $\beta(0) = 0$ and $\beta(\infty) = \pi$ (Fig. 9.4).

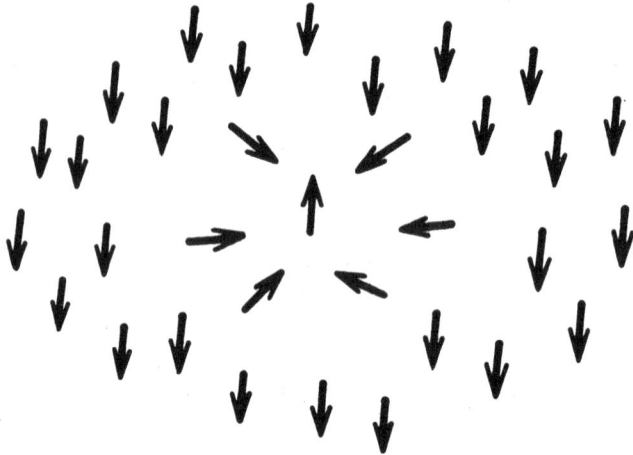

Fig. 9.4. Distribution of the $\hat{d}$ field in a soliton with topological charge $\tilde{m}_d = 1$.

The unusual quantum statistical properties of the solitons are induced by the special topological term in the hydrodynamical action. It exists in addition to the conventional terms in the Lagrangian describing the dynamics of the $\hat{d}$-field,

$$L_{\text{conventional}} = -\frac{1}{2}\frac{\chi_\perp}{\gamma^2}(\partial_t\hat{d})^2 + \frac{1}{2}(\rho_{sp})_{ij}\nabla_i\hat{d}_\alpha\nabla_j\hat{d}_\alpha . \qquad (9.13)$$

This $L_{\text{conventional}}$ may be obtained from the Hamiltonian in Eq. (4.3) by excluding the spin-density variable $\vec{S}$; the dynamical invariant – spin projection $\vec{S} \cdot \hat{d}$ – is taken to be zero. The Lagrangian in Eq. (9.13) completely defines the low frequency dynamics of the $\hat{d}$-field with the condition $\vec{S}\cdot\hat{d} = 0$.

The topological term does not change the motion equations for $\hat{d}$, but defines the quantum statistics, since it describes the change in the wave function of the system in the process of permutation of two identical solitons, or under 2π rotation of a single soliton. The latter property defines the spin of the soliton. This topological term is the so-called Chern-Simons or θ-term in the hydrodynamical action for the $\hat{d}$ field.

The Chern-Simons term is related to a mapping of the three-dimensional space-time (x, y, t) onto the sphere S^2 of the unit vector $\hat{d}$. The classes of such mapping form the homotopy group $\pi_3(R)$, which for our case $R = S^2$ is the group of integers, $\pi_3(S^2) = Z$. The topological invariant, which describes such mapping of 3-dimensional sphere to a 2-dimensional sphere, is known as Hopf invariant. The Hopf invariant cannot be expressed in terms of the $\hat{d}$ field explicitly, but only through the auxilary "gauge" field $\tilde{A}_\mu$ ($\mu = 0, 1, 2$), whose field strength is related to the $\hat{d}$ field:

$$\tilde{F}_{\nu\lambda} = \partial_\nu \tilde{A}_\lambda - \partial_\lambda \tilde{A}_\nu = \hat{d} \cdot \partial_\nu \hat{d} \times \partial_\lambda \hat{d} \ . \tag{9.14a}$$

In terms of the gauge field the Hopf term is

$$H^{\text{Hopf}} = \frac{1}{32\pi^2} \int d^2x \ dt \ e^{\mu\nu\lambda} \tilde{A}_\mu \tilde{F}_{\nu\lambda} \ . \tag{9.14b}$$

The gauge field $\tilde{A}_\mu$ should be expressed in terms of $\hat{d}$ as the solution of Eq. (9.14a). There is some freedom in choosing the solution for $\tilde{A}_\mu$, since it is defined with accuracy of any arbitrary gradient of scalar function. The integral value of H^{Hopf} nevertheless does not depend on this degree of freedom.

The Chern-Simons term in action is proportional to the Hopf invariant:

$$S_\theta = \int d^2x dt L_\theta = \hbar\theta H^{\text{Hopf}} \ , \tag{9.15}$$

and parameter θ is the characteristic of the ground state of the fermionic system.

An example of the nontrivial integer value of the topological charge H^{Hopf} is the process of the time dependent rotation, of the soliton with the topological charge $\tilde{m}_d = 1$ in Eq. (9.12). We describe this as the rotation of the $\hat{d}$ vector field around, say, z axis by angle $\alpha(t)$:

$$\hat{d}_\alpha(\vec{r}, t) = \hat{z}\cos\beta(\rho) + (\hat{\rho}\cos\alpha(t) + \hat{\phi}\sin\alpha(t))\sin\beta(\rho) \ , \tag{9.16}$$

in which $\alpha(t)$ changes from 0 at $t = 0$ to 2π at final time $t = t_0$, so that the solitonic structure returns to its initial configuration at $t = t_0$ (one can take for example $\alpha(t) = 2\pi t/t_0$). This process gives the continuous mapping of space-time (x, y, t) onto the sphere S^2 of the unit vector $\hat{d}$ with the topological charge $H^{\text{Hopf}} = 1$. This can be checked by direct calculation of H^{Hopf} if one chooses some expression for the gauge field A_μ, compatible with Eq. (9.14a). For example

$$A_\mu = (1 - \cos \beta(\rho)) \nabla_\mu (\alpha(t) + \phi) \ .$$

Substituting Eq. (9.16) into the action $S = S_\theta + \int d^2x dt \ L_{\text{conventional}}$, one obtains the change in the action, induced by the process of 2π rotation of soliton $\Delta S = -2\pi^2 \chi_\perp / t_0 \gamma^2 + \hbar\theta + Et_0$ where E is the energy of the soliton. If the rotation is very slow (adiabatical), i.e. $t_0 \to \infty$, then the conventional action acquires the regular form Et_0, while the Chern-Simons term leads to an additional change of the action:

$$\Delta S = \hbar\theta \ . \tag{9.17a}$$

From this equation it follows that the wave function of the soliton, which is proportional to $e^{iS/\hbar}$, is multiplied in the process of 2π rotation by $e^{i\theta}$. On the other hand we know that the spin rotation by arbitrary angle α must result in the change of the wave function of the soliton by the factor $e^{i\alpha s/\hbar}$, where s is the spin of the soliton. So from Eq. (9.17a) it follows that the spin of the soliton is expressed in terms of the parameter θ:

$$s = \hbar\theta/2\pi \ . \tag{9.17b}$$

The same result for the change of the action, $\Delta S = \hbar\theta$, and therefore for the change in the wave function, the multiplication by $e^{i\theta}$, is given by the process of the adiabatical permutation of two identical solitons with topological charges $\tilde{m}_d = 1$. This means that the parameter θ defines also the statistics of the soliton: if $\theta = 2\pi n$, the wave function is invariant under permutation, therefore the soliton obeys the Bose statistics. For $\theta = (2n + 1)\pi$ the wave function changes sign after permutation, therefore with this θ the soliton is a fermion. Thus the quantum statistics of the topological solitons, according to Eq. (9.17), is in correspondence with the conventional relation between spin and statistics: the particles with the half-quantum spin are fermions, and particle with integer spin are bosons.

9.10. *Quantum Statistics of Solitons in the ^{3}He Film*

The important property of the θ term is that the Lagrangian L_θ in Eq. (9.15) is not invariant under the "gauge" transformation

$$A_\mu \rightarrow A_\mu + \partial_\mu\chi , \qquad (9.18)$$

which reflects the freedom in choosing the solution of Eq. (9.14a) for A_μ. Under this transformation

$$L_\theta \rightarrow L_\theta + \frac{\theta}{32\pi^2}e^{\mu\nu\lambda}\tilde{F}_{\nu\lambda}\partial_\mu\chi . \qquad (9.19)$$

This gauge transformation corresponds to spin rotations by an arbitrary angle $\chi(x,y,t)$ about axis $\hat{d}$, which do not change the physical state of the system and therefore should not change the hydrodynamical action. So the total integral should be invariant under gauge transformation, and this results in the following important consequence. If the parameter θ does not depend on the space and time coordinates, integration by parts leads to annulation of the second term in Eq. (9.19) since $e^{\mu\nu\lambda}\partial_\mu\tilde{F}_{\nu\lambda} = 0$. In the opposite case, if θ depends on the space-time coordinates, there is no annulation due to the term $e^{\mu\nu\lambda}\tilde{F}_{\nu\lambda}\partial_\mu\theta$. In this case the action is not gauge invariant and therefore cannot exist.

It follows that the parameter θ should not depend on the external conditions, such as film thickness, otherwise the inhomogeneous external conditions (inhomogeneous thickness $a(x,y)$) would induce the space and time dependence of θ, which violates the gauge invariance. So if the Chern-Simons term exists in a given condensed matter, then the θ parameter should be fundamental for the condensed matter and should remain constant in certain regions of the external parameters, such as film thickness a in ^{3}He-A. This just reminds one of the behavior of the internal (momentum space) topological charge N of the film, and microscopic calculations (calculation of the hydrodynamical action by integrating over the fermionic degrees of freedom) show that θ is actually related to N by the simple expression:

$$\theta = \pi\frac{N}{2} . \qquad (9.20)$$

Since N is an even number, the elementary soliton in the A-phase film may be either fermion or boson; the quantum statistics of soliton depends on the film thickness and changes when one of the transverse levels crosses the Fermi energy (see Fig. 9.5).

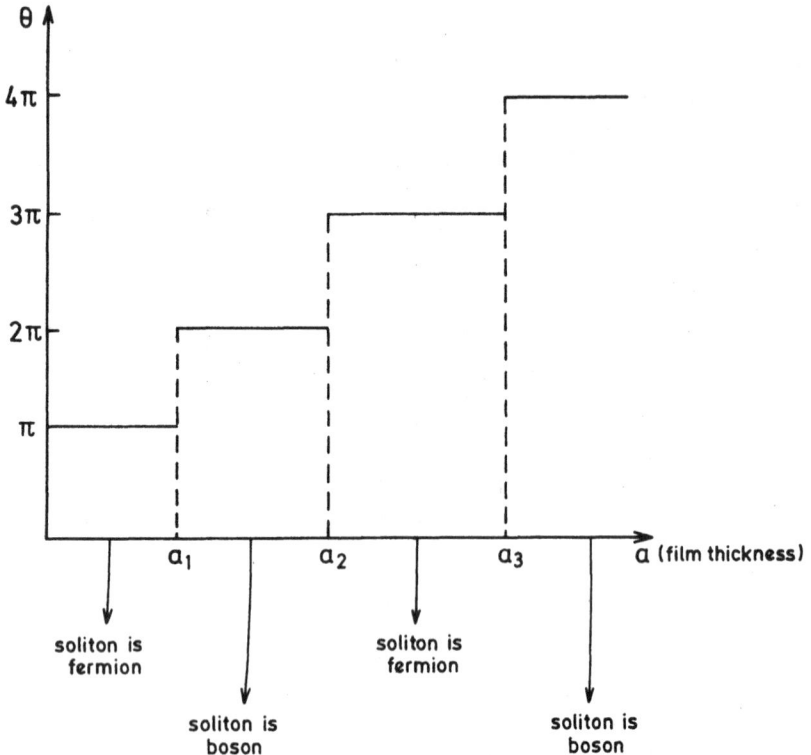

Fig. 9.5. The parameter θ, which defines the quantum statistics of the soliton, is connected with internal topological invariant of the system and changes abruptly, when the film thickness crosses the critical values, where the quasiparticle spectrum touches zero energy level.

9.11. θ-term and Orbital Ferromagnetism

The Hopf invariant in Eq. (9.14) is odd under the time inversion operation $\mathbf{T}H^{\text{Hopf}} = H^{\text{Hopf}}(t \to -t) = -H^{\text{Hopf}}$, and also under orbital rotation by angle π around the x axis: $\mathbf{C}^L_{\pi,x}H^{\text{Hopf}} \equiv H^{\text{Hopf}}(x \to x, y \to -y, z \to -z) = -H^{\text{Hopf}}$. The symmetry $\mathbf{C}^L_{\pi,x}$ corresponds to the two-dimensional space parity; the $\hat{d}$-vector, which is the vector in the spin space, is effected by the orbital rotation only through its coordinate dependence: $\mathbf{C}^L_{\pi,x}\,\hat{d}(\vec{r}) = \hat{d}(\mathbf{C}^L_{\pi,x}\vec{r})$, i.e. $\mathbf{C}^L_{\pi,x}\hat{d}(x,y,z) = \hat{d}(x,-y,-z)$.

The total action however should be invariant under both symmetry operations. This means that the parameter θ, which characterizes the ground state of the system, should be odd under these two discrete symmetries. So in the system the two symmetries must be spontaneously broken in the ground state to have the odd θ.

From these arguments it follows that the θ-term does not exist in conventional spin ferro- or antiferromagnets, where time inversion is not properly broken: for these systems one cannot construct the scalar parameter θ with the required symmetry properties.

Both these discrete symmetries are broken in the A-phase due to the orbital ferromagnetism along the $\hat{l}$ vector, which is fixed along the normal to the film. Due to the orbital ferromagnetism the internal topological charge N is odd: $\mathbf{T}N = -N$, it is also odd under the orbital rotation by angle π around x axis: $\mathbf{C}^L_{\pi,x}N = -N$. This is seen, for example, from its value $N = 2n_0(\hat{l} \cdot \hat{z})$ in the simple case of noninteracting levels.

So the time inversion symmetry does not prohibit the existence of θ term in the hydrodynamic action of the A-phase: both $\theta = N\pi/2$ and H^{Hopf} in Eq. (9.15) change sign under $\mathbf{T}$ and $\mathbf{C}^L_{\pi,x}$ operations, leaving the total action invariant. Therefore the combination of the spin and orbital magnetism is the necessary condition for the existence of the topological term in the quasi-2-dimensional systems.

9.12. *Spin Current QHE in Superfluid ^{3}He-A Films*

The topological quantization of the hydrodynamical parameter θ (parameter in the hydrodynamical action) reminds one of the behavior of the Hall conductivity σ_{xy} in the Quantized Hall Effect (QHE). This is another phenomenon related to the momentum space invariant N. Due to the momentum space topological invariant N in the ^{3}He-A film there exists an analogue of QHE. In the two-dimensional system of electrons the Hall conductivity σ_{xy}, i.e. the response of the electric current to the transverse field, $j_x = \sigma_{xy} E_y$, is quantized, while in an electrically neutral system like the ^{3}He film it is rather the spin current that exhibits quantization in response to external fields.

To relate the spin current QHE in ^{3}He-A to the particle current QHE in a 2-d electron system, we identify the gauge field $\tilde{A}_\mu$ with the electromagnetic field. Then the quantity $\tilde{E}_i = \tilde{F}_{0i}$ plays the part of the electric field, while the quantity $\tilde{j}_i = \delta S/\delta \tilde{A}_i$ plays the part of the electric current. Therefore if one differentiates the θ term, one obtains the quantization of the "Hall" conductivity:

$$\tilde{j}_i = \sigma_{xy} e_{ij} \tilde{E}_j , \quad \sigma_{xy} = \frac{N}{16h} . \tag{9.21}$$

Now one must understand the physical meaning of the quantities $\tilde{j}_i$ and $\tilde{E}_i$. The first variable is related to the spin current. In the general scheme of the linear response on the external gauge fields the spin current may be obtained as response of the system to the fictitious $SU(2)$ gauge field $\vec{A}_\mu$, which is the compensating field for the local spin rotations (in the same manner the mass current in the neutral liquid may be obtained by introducing the fictitious $U(1)$):

$$\vec{j}_\mu = \frac{\hbar}{2} \delta S/\delta \vec{A}_\mu . \tag{9.22}$$

Since the gauge field $\tilde{A}_\mu$ corresponds to rotations about axis $\hat{d}$, it represents one of the components of the $SU(2)$ field $\vec{A}_\mu$. Therefore the "electric" current $\tilde{j}_i$ is just one component of the spin current, this is the current of the spins oriented along the axis $\hat{d}$:

$$\tilde{j}_i = \hat{d} \cdot \vec{j}_i . \tag{9.23}$$

So we obtained the quantization of the spin current response on the special type of external field $\tilde{E}_i$. Now what is $\tilde{E}_i$?

Due to Larmor theorem we know that the real magnetic field, which interacts with the nuclear spins of the ^{3}He atoms, is equivalent to the rotation with constant angular velocity (Larmor frequency). From this theorem it follows that the external field $\tilde{E}_i$, which plays the part of the electric field in this system, proves to be the gradient of the applied magnetic field. As a result the calculations give the following response:

$$\vec{j}_i = -e_{ij}\frac{N\hbar}{32\pi}\gamma\partial_j(\vec{H} - \hat{d}(\vec{H}\cdot\hat{d})) \ . \tag{9.24}$$

Thus the factor in the off-diagonal response of the spin current on the gradient of the magnetic field is quantized in terms of the momentum space topological invariant N.

9.13. *Topological Quantization in Other Superfluid Phases of ^{3}He Films*

In other possible phases in the film, the planar state and the A_1-phase (the latter should exist in high enough magnetic fields), there is also a close relationship between the quantization of the hydrodynamical parameters and that of the characteristics of the solitons, which have the same origin – the existence of the momentum space topological invariant.

In the planar state the time inversion symmetry is not broken. Nevertheless the ground state is described by the topological invariant, which however has a different structure:

$$\tilde{N} = \frac{1}{24\pi^2}\int \mathrm{tr}\tau_3\sigma_z\mathbf{G}\partial\mathbf{G}^{-1}\wedge\mathbf{G}\partial\mathbf{G}^{-1}\wedge\mathbf{G}\partial\mathbf{G}^{-1} \ . \tag{9.25}$$

Here the commonly used notation $\wedge$ stands for the antisymmetric permutation of the (momentum space) derivatives in Eq. (9.10). It differs from Eq. (9.10) by an additional factor $\tau_3\sigma_z$. This invariant is exactly zero for the A-phase, which may be checked by substitution of Eq. (8.20a) for the A-phase Bogoliubov matrix. The planar state Bogoliubov matrix in Eq. (8.20b) results in $\tilde{N}_n = 2$ for each of the transverse levels and the total, $\tilde{N} = 2n_0$, for the n_0 filled levels in the film.

This internal charge leads also to quantization of the physical parameter. For example the 4π nonsingular disclination in the field of the degeneracy parameter – the orthogonal matrix $R_{i\alpha}$ – has, instead of the fractional spin, fractional fermionic charge $\tilde{N}/4$. For the electrically charged system with symmetry of the planar state this means that the electric charge of this topological defect is $e/2$ for $N = 2$. An equivalent situation can occur in magnetic systems, which combine the broken symmetries of the spin and orbital antiferromagnets.

The A_1 phase, which exists in the magnetic field, is the combination of the spin and orbital ferromagnets. It contains the Cooper pairs of fermions with only one spin component, which is antiparallel to the magnetic field direction, while in the A-phase and planar state both species of fermions form Cooper pairs. So the particle supercurrent and spin supercurrent coincide in this phase. The momentum space topological charges $\tilde{N}$ in Eq. (9.25) and N in Eq. (9.10) are equal for the A_1- phase: $\tilde{N} = N = n_0$. As a result the nonsingular disclination, which is simultaneously a 4π nonsingular vortex, has both fractional spin with a magnitude twice less than in the A-phase, $\frac{n_0 \hbar}{4}$, and fractional charge $\frac{n_0}{4}$ with the magnitude twice less than in the planar state. So this vortex should obey the intermediate statistics in some regions of the film thickness.

10
Conclusion

The superfluid phases of ^{3}He, with their fermionic and bosonic vacuum states of highly broken symmetry, appeared to be representative of condensed matter, which has the closest relations to the vacuum in particle physics. A lot of phenomena, which had only been proposed as hypothetical for the particle physics vacuum, are of great importance for these phases. This may be exemplified by the puzzling singularities in low temperature orbital dynamics of the A-phase, which proved to be just of the same origin as the chiral anomaly and zero-charge effect in particle physics. Some more analogies are still waiting to be explored.

One of them is the one-dimensional world of fermions localized in the cores of quantized vortices. In conventional s-wave superconductors the excitations localized near the vortex axis, have a small energy gap $\sim \frac{\Delta^2}{\epsilon_F} \ll \Delta$ as compared with the gap Δ of the delocalized excitation far from the vortex. As a result they play decisive role in the kinetics and thermodynamics of the superconductor at low temperatures.

In modern relativistic field theories the vortices (strings) with the localized fermions are also discussed. It is important that, as distinct from the Abrikosov vortices in conventional superconductor, in the majority of the models for strings one or more branches of fermions are the so-called fermionic zero modes, which means that they are gapless. As a rule the number of the gapless fermions is related to the topological charge of the vortex (its winding number).

In ³He-B due to the complicated structure of the vortex cores with several spontaneously broken symmetries within the core, these vortices also can have fermionic zero modes. These one-dimensional gapless fermions should play the most essential role in the thermodynamic and dynamic properties of the rotating superfluid. These properties should be very closely related to the $1+1$ quantum field theory, since the soft (Goldstone) degrees of freedom of the vortex (translational for the type 1 vortices + rotational for the V2 vortices with broken axial symmetry) play the part of gravity and other bosonic fields.

Also the string theory may be important for the helium-3 phases at low temperature, though the action for the string in particle physics is essentially different from the action for some topological strings in superfluid ³He. While the action for strings in particle physics is proportional to the area of the surface swept by the string during its motion, the action for the quantized vortex is proportional to the volume inside this surface.

Another common problem is the vacuum instability under some external conditions. In the superfluids such phenomenon arises, for example, when the external object is moving with supercritical velocity. When the velocity exceeds the Landau critical velocity, the vacuum proves to be different for the observer, which moves together with the object, and for the observer in the laboratory frame. The vacuum instability in this case results in the creation of excitations by the moving object. A similar situation occurs for the object, moving with acceleration in the vaccum of particle physics. The difference in the vacuum state for two observers leads to the Unruh effect: the accelerated object acquires the effective temperature, which is proportional to the acceleration. In Sec. 6 we have already considered the similar process of particle creation by the accelerated object in the B-phase. However one may expect that in the superfluid phases of ³He, where there are a lot of order parameter fields, interacting with the fermionic vaccum, with some of them playing the part of the gravity field, some more analogies with the Unruh effect and Hawking radiation will appear.

The supersymmetry, symmetry under transformation which mixes fermions and bosons, still has not found its practical applications in particle physics. Can it still be applied for superfluid phases of ^{3}He is not clear. However some indications exist, that this idea can be important for ^{3}He. In the B-phase this follows from the observation, that there is a sum rule for the squares of the gaps (massess) for the bosons and fermions. If Δ_{JT} is the gap of the bosonic collective mode in Eq. (4.27), where $J = 0, 1, 2$ and $T = \pm$ are the quantum numbers, which describe the excitations, then in the BCS weak coupling approximation one has for each J the following relation:

$$\Delta_{J+}^2 + \Delta_{J-}^2 = 4\Delta_B^2 \, , \tag{10.1}$$

where Δ_B is the gap in the fermionic spectrum according to Eq. (5.6). This indicates the hidden symmetry of the BCS Hamiltonian, which possibly has some relation to the supersymmetry at least in the weak-coupling approximation. It is also possible, that the hidden symmetry in the A-phase, which leads to the zero mass of the W-boson (Sec. 5.15), is related to the supersymmetry.

A group of phenomena, related to the disorder, also can be observed in superfluid phases of ^{3}He, for example in the ^{3}He film on a rough substrate. The roughness of the solid surface plays the part of the impurities which tend to destroy the anisotropic order paramater $A_{\alpha i}$. On decreasing the film thickness the impurities become more effective in destroying the Cooper pairing and at some critical thickness the spatial average of the order parameter would disappear, $\langle A_{\alpha i} \rangle = 0$. However the correlators of the order parameter, which in principle are more symmetric than the order parameter itself may still survive such as

$$\langle A_{\alpha i} A_{\alpha i} \rangle \, , \tag{10.2}$$

$$e_{\alpha\beta\gamma} \langle A_{\alpha i} A_{\beta i} \rangle \, , \tag{10.3}$$

$$e_{ijk} \langle A_{\alpha i} A_{\alpha j} \rangle \, , \tag{10.4}$$

$$e_{\alpha\beta\gamma} \langle A_{\alpha i}^* A_{\beta i} \rangle \, , \tag{10.5}$$

$$e_{ijk} \langle A_{\alpha i} A_{\alpha j}^* \rangle \, , \tag{10.6}$$

$$e_{ijk}e_{\alpha\beta\gamma}\langle A^*_{\alpha i}A_{\beta j}\rangle \ . \qquad (10.7)$$

These states still correspond to some long-range order. Equation (10.2) describes the isotropic superfluidity of the four-particle correlated state, since the scalar order parameter $\psi = < A_{\alpha i}A_{\alpha i} >$ acquires under the $U(1)$ gauge transformation the factor 4, i.e. $\psi \to \psi e^{4i\alpha}$, which means that the elementary boson in the Bose-condensate contains 4 fermionic particles. Equations (10.3) and (10.4) describe the superfluid four-particle correlated state with correspondingly spin and orbital anisotropy; the former is thus the superfluid ferro- or antiferromagnet, while the latter is the superfluid liquid crystal. The correlators in Eqs. (10.5-7) correspond to the nonsuperfluid states with the magnetic or/and orbital properties.

When these correlators vanish, the more complicated many-particle correlators may arise. So with decreasing of the helium-3 film thickness one may expect the hierarchy either of the thermodynamic phase transitions or Lifshitz transitions into the exotic many-particle correlated states, including spin-glass and superfluid-glass states. Some of the many-particle correlated states can have internal topology, which leads to the parameter quantization or to the phenomena, similar to the fractional quantum Hall effect.

However it seems that the most important phenomena in the superfluid ³He phases are related to quantum coherence at very low temperature. Note that far from the container surfaces, well deep in the bulk liquid, these ordered liquids are essentially pure systems. This allows in future to utilize the effects of the quantum coherence in a large volume, which may be important in particular for possible modifications of the foundations of quantum mechanics.

Acknowledgements

This text was written during my visit to the *Centre de Recherches sur les Tres Basses Temperatures* in Grenoble and was finished at *NORDITA Institute*, Copenhagen. I am indebted to J. Flouquet, P. Monceau, R. Rammal, N. Schopohl, P. Nozieres, and A. Luther for their kind organization and scientific help during my work on the manuscript.

References

For reviews on the superfluid phases of ^{3}He, see:

D. Vollhardt and P. Wölfle, *The Superfluid Phases of Helium 3* (Taylor and Francis, 1990);

A. J. Leggett, *Rev. Mod. Phys.* **47**, 331 (1975);

J. C. Wheatley, *Rev. Mod. Phys.* **47**, 415 (1975);

P. W. Anderson and W. F. Brinkman, in *The Physics of Liquid and Solid Helium*, Part II, edited by K. H. Bennemann and J. B. Ketterson (Wiley, New York, 1978), Chapter 3;

D. M. Lee and R. C. Richardson, *ibid.*, Chapter 4;

G. E. Volovik, "Symmetry in Superfluid ^{3}He", in *Modern Problems in Condensed Matter Sciences (Helium Three)*, edited by W. P. Halperin and L. P. Pitaevskii (North-Holland, 1990), Chapter 2.

For reviews on the superfluid properties of ^{3}He-A, see:

G. E. Volovik, *Usp. Fiz. Nauk* **143**, 73 (1984) [*Sov. Phys. Usp.* **27**, 363 (1984)];

H. E. Hall and J. R. Hook, 1986, "The hydrodynamics of superfluid ^{3}He", in *Progress in Low Temperature Physics Vol. IX*, edited by D. F. Brewer (North-Holland, 1986).

For reviews on topological defects in condensed matter, see:

N. D. Mermin, *Rev. Mod. Phys.* **51**, 591 (1979);

V. P. Mineev, "Topologically stable defects and solitons in ordered media" in *Soviet Scientific Reviews* A2, edited by I. M. Khalatnikov (Harwood Academic Publishers, 1980), p. 173;

M. Kléman, *Points, Lines and Walls in Liquid Crystals, Magnetic Systems and Various Ordered Media* (Wiley, New York, 1983);

L. Michel, "Symmetry defects and broken symmetry. Configurations. Hidden symmetry", *Rev. Mod. Phys.* **52**, 617 (1980).

For reviews on quantized vortices in ^3He, see:

M. M. Salomaa and G. E. Volovik, *Rev. Mod. Phys.* **59**, 533 (1987);

P. J. Hakonen, O. V. Lounasmaa, and J. Simola, *Physica* B160, 1 (1989);

G. A. Kharadze, in *Modern Problems in Condensed Matter Sciences (Helium Three)*, edited by W. P. Halperin and L. P. Pitaevskii (North-Holland, 1990);

A. L. Fetter, 1985b, "Vortices in rotating superfluid ^3He", in *Progress in Low Temperature Physics Vol. X*, edited by D. F. Brewer (North-Holland, 1987);

V. P. Mineev, M. M. Salomaa, and O. V. Lounasmaa, "Superfluid ^3He in rotation", *Nature* **324**, 333 (1986).

On monopoles and boojums in ^3He-A, see:

N. D. Mermin, "Surface singularities and superflow in ^3He–A", in *Quantum Fluids and Solids*, edited by S. B. Trickey, E. D. Adams, and J. W. Dufty (Plenum, New York, 1977), p. 3;

M. M. Salomaa, "Monopoles in the rotating superfluid helium-3 A-B interface", *Nature* **326**, 367 (1987);

G. E. Volovik, "Defects on the boundary between A and B phases of superfluid ^3He", *Pis'ma ZhETF*, **51**, 396 [*JETP Lett.* **51**, 449 (1990)].

On collective modes in superfluid phases of ^3He, see:

J. M. Kyynäräinen, K. Torizuka, A. J. Manninen, A. V. Babkin, R. H.

Salmelin, J. P. Pekola, and G. K. Tvalashvili, "Ultrasonic spectroscopy of the real squashing mode in rotating ^{3}He-B", *ZhETF* **98**, 516 (1990) [*Sov. Phys. JETP* **71**, 287 (1990)];

W. P. Halperin and E. Varoquaux, "Order-parameter collective modes in superfluid ^{3}He", in *Modern Problems in Condensed Matter Sciences (Helium Three)*, edited by W. P. Halperin and L. P. Pitaevskii (North-Holland, 1990), Chapter 7;

R. H. McKenzie and J. A. Sauls "Collective modes and nonlinear acoustics in superfluid ^{3}He-B", in *Modern Problems in Condensed Matter Sciences (Helium Three)*, edited by W. P. Halperin and L. P. Pitaevskii (North-Holland, 1990), Chapter 5.

On the broken relative symmetry in superfluid ^{3}He, see:

M. Liu, "Broken relative symmetry and the hydrodynamics of superfluid ^{3}He", *Physica* **109 & 110 B**, 1615 (1982).

On the A-B interface and other interfaces in superfluid ^{3}He, see:

N. Schopohl and D. Waxman, "Quasiclassical theory of the A-B phase boundary of superfluid ^{3}IIe", *Physica* **B169**, 165 (1991);

M. M. Salomaa and G. E. Volovik, "Cosmic-like domain walls in superfluid ^{3}He-B: Instantons and diabolical points in $(\vec{k}, \vec{r})$ space," *Phys. Rev. B* **37**, 9298 (1988);

T. Sh. Misirpashaev, "Topological classification of defects at a phase interface", *ZhETF* **99**, 1741 (1991) [*Sov. Phys. JETP* **72**, 973 (1991)].

On solitons in superfluid ^{3}He, see:

K. Maki, "Solitons in superfluid ^{3}He" in *Solitons*, edited by S. Trullinger, V. E. Zakharov and V. L. Pokrovskii, (North- Holland, 1986), p. 435.

On the structure of ^{3}He-B vortices with broken axial symmetry in the core, see:

E. V. Thuneberg, "Identification of vortices in superfluid ^{3}He-B", *Phys. Rev. Lett.* **56**, 359 (1986);

M. M. Salomaa and G. E. Volovik, "Topological transition of v-vortex core matter in superfluid ^{3}He-B", *Europhys. Lett.* **2**, 781 (1986);

G. E. Volovik, "Half quantum vortices in the B phase of superfluid ^{3}He", *Pis'ma ZhETF* **52**, 972 (1990) [*JETP Lett.* **52**, 358, (1990)].

On the observation of the ^{3}He-B vortices with broken axial symmetry in the core, see:

J. S. Korhonen, Y. Kondo, M. Krusius, V. V. Dmitriev, Y. M. Mukharskiy, E. B. Sonin, and G.E. Volovik, *Phys. Rev. Lett.* **67**, 81 (1991).

On the observation of the soliton terminating on the linear defect, see:

Y. Kondo, J. S. Korhonen, Y. Kondo, M. Krusius, V. V. Dmitriev, E. V. Thuneberg, and G. E. Volovik, "Nucleation of vortices at the A-B phase boundary in superffluid ^{3}He: nonconservation of circulation and vortices with soliton tails in the B-phase", Report TKK-F-A688, Helsinki University of Technology, Otanieni (1991).

On the observation of the topological transition in the ^{3}He-A vortices, see:

J. P. Pekola, K. Torizuka, A. J. Manninen , J. M. Kyynäräinen and G.E. Volovik, "Observation of a topological transition in the ^{3}He-A vortices", *Phys. Rev. Lett.* **65**, 3293 (1991). Completely singularity free A-phase vortex texture in a rotating vessel (without boojums) with compensating surface layer of antivortices is discussed in: K. Torizuka, J. P. Pekola, A. J. Manninen, and G. E. Volovik, *Pis'ma ZhETF* **53**, 263 (1991) (in English).

For the spin and statistics of solitons in ^{3}He films and Quantum Hall Effect, see:

G. E. Volovik and V. M. Yakovenko, *J. Phys.: Condens. Matter* **1**, 5263 (1989);

G. E. Volovik, "Analog of quantum Hall effect in superfluid ^{3}He film," *ZhETF* **94**, 123 (1988) [*Sov. Phys. JETP* **67**, 1804 (1988)].

For the chiral anomaly in ^{3}He-A, see:

G. E. Volovik, *ZhETF* **92**, 2116 (1987) [*Sov. Phys. JETP* **65**, 1193 (1987)];

M. Stone and F. Gaitan, "Topological charge and chiral anomalies in fermi superfluids", *Annals of Phys.* **178**, 89 (1987).

On the analogy between quantum field theory in particle physics and in ^{3}He, see:

G. E. Volovik, *J. Low Temp. Phys* **67**, 301 (1987);

G. E. Volovik, "Superfluid ^{3}He-B and gravity," *Physica* **B162**, 222 (1990).

On the dynamics of the Goldstone fields in superfluid ^{3}He, see:

W. F. Brinkman and M. C. Cross, "Spin and orbital dynamics of superfluid ^{3}He", in *Progress in Low Temperature Physics, Vol. VIIA*, edited by D. F. Brewer (North-Holland, 1978), p. 106.

On the possible supersymmetry in superfluid ^{3}He, see:

Y. Nambu, "Fermion-boson relations in the BCS-type theories", *Physica* **D15**, 147 (1985).

On the fermionic zero modes in ^{3}He vortices, see:

G. E. Volovik, "Localized fermions on quantized vortices in superfluid ^{3}He-B", *J. Phys.: Condens. Matter* **3**, 357-368 (1991).

On the pair creation in ^{3}He and Unruh effect, see:

N. Schopohl and G.E. Volovik, "Schwinger pair production in the orbital dynamics of ^{3}He-B", submitted to *Annals of Physics*.

On the quantum phase slippage and vortex nucleation in ^{3}He, see:

N. B. Kopnin, P. I. Soininen, and M. M. Salomaa, "Dynamics of vortex nucleation in ^{3}He flow", *Phys. Rev.* **B44** (1991).

www.ingramcontent.com/pod-product-compliance
Lightning Source LLC
Chambersburg PA
CBHW050639190326
41458CB00008B/2340